The serial time-zoned multiverse as described in this book is featured on the front cover as a closed loop of universes which move clockwise around the loop; expanding then contracting as they do so. The multiverse exhibits the state of eternality (eternalism in philosophy). The central image helps explain this. It depicts three structures built near Dover castle England at different times—yet they are intimately connected: the Roman lighthouse at right constructed ca AD130; St Mary de Castro church constructed ca AD 1000; and at left a church tower restored between ca AD 1780 and 1888. This complex of structures represents activity (events) which span a duration of at least 1758 years. To observers who confine their thinking inside this universe these structures are merely relics of our past slowly crumbling away. To observers outside the multiverse they see each of these structures being built in many universes—at the same multiverse cosmic time.

Photo by Gerald Holdsworth 2009.

TIME

and the

MULTIVERSE

The Foundations of Human Existence

GERALD HOLDSWORTH PhD

ARCHWAY
PUBLISHING

Archway Publishing books may be ordered through booksellers or by contacting:

Archway Publishing
1663 Liberty Drive
Bloomington, IN 47403
www.archwaypublishing.com
1 (888) 242-5904

ISBN: 978-1-4808-4979-2 (sc)
ISBN: 978-1-4808-4978-5 (hc)
ISBN: 978-1-4808-4980-8 (e)

Library of Congress Control Number: 2017913900

Print information available on the last page.

Archway Publishing rev. date: 11/29/2017

CONTENTS

PREFACE

In late 2005, I had nearly finished the first of several planned interviews gathering information for a biography of Walter Wood, an independently wealthy and influential American explorer, surveyor, and mountaineer who led expeditions into the remoter parts of the Yukon from 1935 to 1951. This interview was with ninety-four-year-old Robert (Bob) Hicks Bates in Exeter, New Hampshire. He had been a prominent member of three of Walter Abbott Wood's Saint Elias Mountains expeditions.

He recalled details of the last expedition of Project Snow Cornice in 1951. We got to the part about the loss of the expedition Norseman airplane on July 27, 1951. Bob told me something about Walter's wife, Foresta, who had died in the plane crash. I continued taping and listened in mild shock while Bob told me that Foresta had known through several psychic readings over many years including one just the year before (1950) about her predicted early death in an accident.

An earlier incident had occurred on that expedition while some members of the group were being flown to a base camp on the Seward Icefield. The engine began to miss some beats while they were over the dangerously crevassed Seward Outlet Glacier. Only pilot Maurice King knew it was time to pull a lever to switch over to the reserve fuel tank. This was an impressive place to do it provided your nerves held up! That night after supper at base camp, Foresta told the group she thought that incident had been *it* for her, and she told them straight-forwardly about her predicted demise.

Walter did not believe in any of this prediction business. Regardless, it was now significant for me to learn more about Foresta. I began to gain the impression that Foresta had been an integral part of Walter's operations. I planned a double biography and to start the story at the end of August 1951, right after the closing days of the futile, month-long air search that followed before flashing back to their early lives as some movie scripts have dramatically done.

However, that night, I couldn't ignore some powerful and determined thoughts. It had to do with writing into the story an event that featured the enigmatic phenomenon called *prescience*. This grey area includes precognitions and premonitions. I didn't know how it worked, and as I subsequently discovered, no one else did either. I cannot write about phenomena I don't understand. But first, I needed to check out Bob's story.

That checking out was forthcoming. In the meantime, I needed to wait three more years to clear myself from committed field research in east Pacific climate change. Meanwhile, Bob died in 2007. Then a visit to New York City in the fall of 2008 brought me in contact with Foresta's niece, installation artist Lucy Hodgson, who told me that yes indeed Foresta had discussed her visits to psychics with her family as well as many other people just as if it constituted normal chitchat.

It started in the 1920s after Foresta survived two near-fatal taxi accidents in New York. The prognosis was that the third accident would be fatal. Her sister, Daphne, joined in on this inquiry and was told she would be involved in a road accident. In 1978, Daphne was involved in a serious collision when another vehicle hit the car in which she was a passenger. Her husband, who was driving, was killed instantly, and she died in the hospital from her injuries.

On a second visit to New York a year later, I visited the Wood side of the family and learned that Foresta had been in possession of a life chart (I assume an astrological one) and that the psychic had pointed to an accidental end to Foresta's life. Further inquiry on the matter met a dead end, and the matter lapsed while I studied the evidence supporting prescience.

I joined the Victoria chapter of the Questers, a group that used to sponsor lectures on topics that included preternatural phenomena. I had by then synthesized a two-dimensional diagram that described a system of universes that existed simultaneously. It fitted the system philosophers called eternalism. This state can also be referred to as eternality, but there was no physics or any diagrams that supported it. There was usually no mention of any data that would support eternality. There was no mention of stories like the one I recounted about Foresta. Meanwhile I had acquired John Dunne's now classic book *An Experiment with Time* and learned about precognitive dreams. I saw no reason to ignore this evidence; I considered it as a window to a much expanded cosmos: a multiverse.

Soon after that, I received an email from a Mrs. Nuna MacDonald in Maine; she had read an article in the *New York Times* about the current warming climate resulting in the recession of glaciers in many parts of the world. Concurrent with this trend, mountaineers were finding long-lost aircraft and human remains that were beginning to melt out of their icy tombs. In that article, I had been quoted as saying I was looking for the wreck of an airplane near the Yukon-Alaska border. My name was already in the media as being linked to climate change research in regions of high mountain glaciers, and it would also have been linked to the Arctic Institute of North America in Calgary which was my home base at the time.

That was how Nuna was able to contact me. She was the daughter of Dr. A. Lincoln Washburn, an arctic soils geomorphologist who in 1951 was director of the Arctic Institute in Montreal. Nuna told me that her father had flown to Yakutat to help in the search for the lost aircraft. I asked her if she knew if her father had kept a diary during the search. It turned out that her brother, Land Washburn, in Redmond, Washington, had his diaries. I soon had a copy of the entries for August 6 to September 3, 1951; they turned up gold.

Al Luke, a helicopter pilot (later inducted into the Illinois Aviation

Hall of Fame) was at Yakutat with a Bell helicopter during the search. He must have talked to Foresta before she flew in to the base camp on a nunatak protruding above the Seward Glacier. She had told him about the last prediction. The ALW diary August 30 reads,

> Al Luke discoursed on his philosophy of life last night. He appears to believe in clairvoyance. Foresta had told him [Luke] that a fortune teller in Europe last summer had told her she would be killed in a crash within two years and she spoke about it again to the boys at the nunatak.

Bob Bates, my other source, had told me that the year before, 1950, a psychic (likely in Paris, where she had studied sculpture at Fontainebleau in the 1920s) had told her she would die in an accident in twelve to eighteen months. The plane crash fits into that time slot.

What comes out of that story is quite profound as well as instructive. It is a direct statement that some people accept their lives as being predestined and even after being repeatedly told a disaster is looming, they do not or apparently cannot waver from a perceived path in life.

In this case, Foresta's life was principally being determined by her husband, Walter. That reasoning extends to all the other people on the expedition especially to the pilot of the ill-fated expedition airplane that had just had an engine change earlier that month.

But as I will show, it turns out that everyone's life in this universe is being determined by processes beyond his or her control. This is demonstrated in chapter 2 by two geometric diagrams that replace John Dunne's diagrams of 1927. In chapters 3 and 5, I discuss several well-documented cases in which this situation prevails. The cases involve people who are deemed normal, rational individuals. For example, I relate the convincing and best-documented story of W. T. Stead, who drowned in the sinking of the *Titanic*. This happened after he had repeatedly received warnings of a disaster given to him well ahead of the event. He even had his own premonitions of danger from "crowds and water." Yet he did not take any of the warnings seriously enough. He thought that the crowds would be a mob that attacked him for his radical investigative journalism covering social, corporate and political issues of the day. The facts about the sinking of the *Titanic* in this universe showed that Stead's key words were correct but his interpretation of the forebodings were incorrect. Stead's case alone was enough to set me up for the multiverse project. I document many others in chapter 5.

This leads to the procedural aspects taken for starting the project. The simple-enough questions stemming from my earlier comments were these: Can a mechanism be found whereby some people can see the future? How can a life in this universe be predetermined? Where might information about the future exist? If we are to have faith in our grasp of physics and cosmological principles it must exist outside this universe. The task: How can we structure a multiverse so that it broadly accommodates precognition?

Using geometry and logic, I constructed the multiverse Blueprint and one highly likely interpretation of it that I call the Cosmological model. All I needed to do, and could do, was to verify it with as much good quality data as possible. As every researcher knows, even if you have only a shred of a hypothesis, the amount of those data which are consistent with that hypothesis will extend the life of that hypothesis. Throughout the book I use the term Cosmological model to refer to the complete multiverse structure that I derive from the Blueprint. Elsewhere I use the terms cosmology or cosmological only where parts of it are being discussed or where I mention that other interpretations of the Blueprint are possible despite their being unlikely. The Blueprint, a crucial two dimensional line drawing, corresponds in principle to John Dunne's primitive model of time and is successful in passing many of the tests applied to it. Other tests are made on the Cosmological model where the Blueprint cannot be used because such information is hidden.

It was easy to see that no detailed mechanism to account for transmission of information over vast, cosmic distances was likely to be forthcoming in my lifetime. I decided to continue regardless and rely on the strength of the evidence for precognition and premonition. This showed the very compelling, overall, large-scale performance of the multiverse model presented. However, there was a conundrum. Whereas I was to discover that information must travel in the multiverse counter to the direction of movement of the universes, in precognition cases, information (e.g., in the form of images) must travel much faster.

As planned, during this exercise, I hadn't looked for possible hints from books or on the Internet other than what I may have remembered from earlier interrogations of various non mathematical science books. I was aware of terms such as *block time*, a closed-universe solution to the Einstein field equations, that there is no distinction between past present and future (attributed to Einstein), and even "Time exists in order to stop everything happening at once!" That seems a good quip, but it is actually contrary to what you will find out in chapter 2.

Until 2008 I had never heard of John William Dunne, a British army officer, aircraft-designer/engineer, and test pilot with a sidespin activity that involved figuring out from his frequent precognitive dreams how the future might be revealed to us. He reached his solution of the time problem in the mid-1920s after working long hours at his drafting board (just as I did) and producing a strange-looking geometrical diagram that few if any persons could understand. This diagram is in his 1927 book *An Experiment with Time,* which I discovered in the fall of 2008. It is best described as an experiment that went off the rails.

Later that year, I began to systematically test the model of the multiverse for its predictive abilities. I began to rely on Internet resources to strengthen my claims and quickly became a victim of google-itis. It was actually the Internet that helped me through. I began to show the symptoms of "a time-haunted man," a phrase used by John B. Priestley, whose thoughtful and colorful book *Man and Time* I acquired in 2009. My book list expanded rapidly to include books of very great vintage. Historical figures such as Alfred North Whitehead,

Arthur Eddington, Albert Einstein, Wolfgang Pauli, and others all the way up to modern writers of new age physics books devoted entirely to the subject of time formed a seemingly unending list.

Very soon, I had close to a hundred books of this genre. I even procured Agatha Christie's autobiography because it contained a wonderful, unique paragraph about Dunne's book an acquaintance in Baghdad had loaned her. It must have referred to the first (dream) section of the book because no reviewer of that book and certainly not a crime mystery novelist has ever demonstrated to me a complete understanding of the second half of Dunne's book. This is where Dunne lost his connection with reality by drifting into other dimensions without anywhere specifying where they existed. He had prescribed automatic failure in his task by having firmly pronounced in the preface to his book that he had no use for additional universes.

By 2010, I realized I had taken on an immense task and all the draft chapters needed an immediate and serious rewrite. Otherwise, considering the steady growth of ideas in cosmology, I might just get edged out by some other attempt at the same quest. Mark Twain had a theory that unusual clusters of authors writing about the same subject at about the same time was the result of telepathy; he even joined the American Society for Psychical Research to find out more about human abilities to connect to others. I had learned from reading Carl Jung's *Memories, Dreams and Reflections* that there was such a concept as the collective unconscious though this was different. It sounded to me analogous to an immense, natural, Internet-like resource.

The manuscript started with the provisional title of *Discovering a Multiverse*; that morphed into *Exploring a Multiverse* followed by *Exploring the Multiverse, Man and the Multiverse* ... the list grew. It was illustrated with geometrical diagrams based on what I have referred to as the Blueprint and the Cosmological model, which are derived from it. This described a serial time-zoned multiverse (STZM) arranged in a closed loop. Only about 9 percent of the book is taken up with these details; the remainder of the book concerns the interpretation and verification of the model in its many aspects.

I now realized that I was involved in a marathon writing session; as I wrote new interpretations began occurring to me as well as a paradox. I struggled to maintain headway. I had to reread parts of Dunne's book in the geometry section to show exactly where he had gone awry. I revisited David Bohm's book *Wholeness and the Implicate Order* many times. This became arduous because he employed an obtuse style of writing using words of his own creation. There was yet again no mention of a multiverse. Then I realized in 2013 that I needed an editor to find out if I had written the manuscript in the right style or not. To take a break from the geometric formalism that had crept in (much as it had with Dunne), I wrote chapter 4, which gives thumbnail sketches of many interesting and well-known characters whose lives overlapped Dunne's and may have had some influence on his very singular mode of thinking.

I had earlier expectantly checked the Internet to see if there was anyone developing such a new creation as a STZM. By 2014, I was producing draft chapters which were being edited as I progressed. I had then stopped interrogating the Internet for new developments as this activity was holding me up.

Unknown to me, in October, a journal article was published that had a significant influence on the presentation of the new multiverse model. In early 2015, my editor alerted me to this new theory called the many interacting worlds (MIW) hypothesis modeled on a new approach to quantum mechanics. Because of this, substantial parts of chapters 5 and 6 had to be rewritten and thus reedited.

The role of time had already taken on a new twist with the production of the Blueprint which describes a general cosmology. In attempting to geometrically present a concept of the Now moment, dual use of the Blueprint and the special cosmology derived from it took on a new significance. This resulted in another change in the title of the manuscript to *Time and the Multiverse*. I added the subtitle, which the reader may think rather curious, at the suggestion of the Archway Publishing team. It refers to the inside story, which will resonate only with people who have experienced the preternatural.

Here are four positive results provided by the STZM model.

1. As seen in the cosmological model, the multiverse as presented in this book broadly supports the existence of precognition, but lacks specific details.
2. There are two distinct but mutually compatible types of time: one being a timing system ubiquitous throughout the multiverse, and the other being the human-devised clock time with its astronomically based underpinnings.
3. The STZM appears to be complementary to the theoretical quantum mechanical based multiverse (MIW).
4. Using a spiritual link, the purpose of the multiverse became evident: it provides a vastly expanded stage for incarnation of spirits (to become souls) compared with a single universe, which is inadequate in this area.

Moreover, our universe needs to be a member of a multiverse in order to function in a way suggested by the extensive data that continually draws our attention.

However, other accommodation for souls in our own universe may exist. According to NASA, our universe may contain 2×10^{22} (20 sextillion) planets; thus, there is a possibility that some may contain planets similar to ours with beings similar to us. This potentially expands the number of humanoid bodies available to spirits of the type we are familiar with. It is speculation on a large scale and a very inefficient way of providing extra bodies living the same life for soul transmigration. In contrast, the STZM model is very efficient and relatively compact at the largest scale.

Gerald Holdsworth
Cobble Hill, BC. March 19, 2017

CHAPTER 1

Introduction

Nothing ventured, nothing gained.
—Benjamin Franklin (1706–1790)

The True Beginnings

Millennia ago, humans were already starting the quest to determine where they fitted into the grand scheme displayed by the relative movement of the sun, moon, and the myriad of other celestial objects.

By the Middle Ages, the renowned intellectual and polymath Albert the Great (ca. 1200–1280) had asked, "Do there exist many worlds, or is there but a single world? This is one of the most noble and exalted questions in the study of Nature."[1] Can we assume he was thinking of worlds completely outside our universe such as those in the many worlds interpretation or the many interacting worlds approach of quantum physics? It depends on the definition of the word *world* during Albert's lifetime. I take the modern use of worlds as meaning universes.

Three centuries after Albert the Great's great thought, a telescope swung skyward and enabled Galileo Galilei (1564–1642) to examine the orbits of the planets in our solar system as did his more mathematical contemporary, Johann Kepler (1571–1630) who in a sense helped set up Isaac Newton (1642–1727) on his way to great fame.

Newton followed through by establishing a new, gravitationally driven mechanics that allowed him to approximate the orbits of planets as well as the motion of terrestrial objects. For three centuries, his theory was near enough for most purposes, and it is still accurate enough for most of the many types of scientists and engineers practicing today.

The scientific revolution Newton helped start received a major boost in the early decades of the twentieth century by Albert Einstein. However, Newtonian mechanics was by no means replaced by Einstein's new gravity theory, which is contained in the theory of general relativity. Fortunately, this issue doesn't affect anything found in chapter 2.

Hermann Minkowski, Einstein's mathematics professor at ETH (Zürich), introduced a geometric view of space and time. Nevertheless, I do not go along with the mathematical treatment of converting clock time to units of a space dimension even though it gives the seemingly correct answers. However, it is interesting to know that even before Einstein's theory of special relativity was published in 1905, there had already been planted the seed from which quantum theory (QT) was developed. Experimental research by Max Planck and related theoretical work by Einstein indicated discrete, jump-like behavior occurring in classical particle/electromagnetic wave theory. The development of QT mushroomed over several decades well into the twentieth century and is still being reworked in the twenty-first century. I will now espouse my connection with QT, which can generally be equated with the terms *quantum mechanics* and *quantum physics* in a book of this type without formal equations.

Here, I have avoided getting involved in the unintuitive aspects of quantum mechanics and especially in the operation of primitive quantum computers—details about which can easily be found on the Internet. Though I have advocated for them as being essential to the serial time-zoned multiverse (STZM), I am not equipped to deal with them any more than what I have read in books for the general reader in which one finds for example that a mere thimbleful of chloroform will suffice for your first quantum computer setup. But the real issue is exactly what out there computes and delivers the next state of the universe. For that matter, how does a cosmic quantum computer drive the whole multiverse? We may never know. Actually, do we really need to know? It is more important to know how our hearts beat. That, as well as our whole physical body, is updated every Now moment which is defined in chapter 2. The update is like a finite micro-event and is driven by the Cosmic clock corresponding to the multiverse timing system.

In clarification of this and in preparation for later chapters, I use the uppercase C when referring to activities that involve the quantum computer that drives the multiverse and lowercase c when referring to any related activities inside a universe or just a small group of them. Some authors involved in quantum computer research refer to individual universes as having a computer dedicated to each one. My present concept is that a single Cosmic quantum computer might be able to manage all the tasks alone. I have left the matter open for discussion. This introduces one of my main concerns when it comes to the issue of addressing the mechanisms involved in a particular process, the main one being the copying of the Now moment in the direction counter to the forward movement of all the universes around the torus shaped loop of the multiverse.

My main concern is that I am sometimes drawn into an attempt to visualize aspects of multiverse mechanics and the software needed to drive it—and indeed there are some micromechanics in there too. In such cases, I can only think in terms of current digital computing in the area of simulations where continuum equations must be converted to incremental, step-like finite difference equations that form the basis for writing computer code.

Readers who are quantum computer–ready must forgive me for writing as if these are straightforward analogies. This occurs for example when I describe the COPI routine either as a cut-out and paste-in or a copy-and-paste procedure and the display of the new Now moment as well as the simultaneous stepping forward of all the universes inside the torus-shaped field. Remember what Einstein said: "God does not care about our mathematical difficulties. He integrates empirically." He was referring to what I just wrote in the previous paragraph. In chapter 2, you will be introduced to the widely used term *Now moment* or just *Now* (as a noun) always in uppercase.

The general idea that a cosmic quantum computer drives our universe had been thought of years before I even considered writing this book. David Deutsch at Oxford University had thought of trying to connect to other universes with a suitably programmed quantum computer. His logic was that if this experiment turned out to be successful, it would be considered proof that a multiverse exists. This brings up again my earlier questioning of whether computers are in fact in each universe. Deutsch considered that a multiverse existed for reasons other than mine.

After finishing the construction of the STZM and verifying it, I became convinced that it must exist. Of the approximately 106 pages devoted to developing the model and verifying it, approximately 86 percent of those pages are devoted to testing the model and verifying its robustness.

While presenting the evidence, I became acutely aware of a serious problem concerning the definition of time; it was constantly tripping me up as it has been doing to others for millennia. The answer to this problem lies in chapter 2 in the form of the Blueprint diagram. This shows explicitly how the STZM must contain a timing system that has been in existence before our universe's genesis event, denoted in all relevant diagrams as GE. This event replaces the Big Bang which I prefer to circumvent in this book by proposing a non-original alternative.

Substantial parts of the book depend on quantum computer theory, which has already been thought of as applying in this universe.[2, 3] This application now needs to be extended to the multiverse. Physically, this is beyond my knowledge, but in principle, it is straightforward to see that the multiverse Cosmic computer clock needs only be the equivalent of a constant frequency crystal oscillator with a counter just as in our digital computers or by analogy with them.

Such an arrangement is necessary to produce the Cosmic time pips (TP) that have a frequency of $1/\Delta t$, where Δt is the duration between TPs. This leads to a geometrical concept of the finite Now of duration a small amount less than Δt if not equal to it as shown in chapter 5. So this forms the basis of a type of time quite different from our formal time, which is based on astronomical methods[4] and is sliced down from a mean year or a mean day to seconds and used throughout the human community.

In chapter 3, I lay out some of the main evidence supporting the multiverse model as shown by its ability to relate to and solve well-known or easily accessible accounts that deal with such things as the known human ability to perceive future events that eventually actualize in this universe, certain illusions, and paradoxes. The current, largely philosophical subject of why there is a universe at all is automatically carried a logical step further into the multiverse, where answers or responses to this and many more questions become much easier to deal with.

Chapter 4 offers a perspective on the erratic historical development of many of the ideas in chapter 3. To do that, I have assembled a group of generally well-known individuals who contributed to progress in understanding concepts dealing with the possibility of other worlds, the physical and spiritual (nonphysical) sides of life, and the confusing situation surrounding our understanding of the nature of time especially the views of time that have so far existed largely in the realms of philosophy and psychology.

The common characteristic of these three concepts is that they are not amenable to the quantitative methods of established science. Therefore, my list of individuals doesn't include many whom you might otherwise think should be there. I have deliberately selected people who were contemporaneous with or whose lives and thoughts overlapped the life of Dunne, who rates considerable coverage. He used similar geometrical methods[5] employed by me, and if he had not steadfastly refused to believe in multiple universes, he might have succeeded in reaching his goal in 1927 and saved me from writing this book.

Chapter 5 contains extensions of some of the topics covered in chapter 3 as that chapter was beginning to overflow. Instead, I moved the problem elsewhere! Chapter 5 contains recently published results I consider to be highly complementary to the STZM model. In particular, the new model based on an MIW approach is formulated on a new view of quantum mechanics taking into account specific properties reflected in the physical multiverse. Appearing in October 2014, after the manuscript editing of *Time and the Multiverse* had been completed, it received special attention here. In another section is seen a possible geometrical representation of the philosophical terms *being* and *becoming* as the result of defining the finite nature of the Now moment.

Chapter 6 contains a summary with conclusions in the form of thirteen essays I think contain the most essential material. Some individuals could read it after chapter 2 to get a relatively quick grasp of the main aspects of the remainder of the book. The listed topics are not in any particular order; rather, they more or less follow the developments in the book. If any quantum physicists have reached this sentence, I invite them to read the fourth essay first.

Notes

1   http://historymedren.about.com/od/albertusmagnus/a/Albertus-Magnus-Quotes.htm.

2   J. Brown, *The Quest for the Quantum Computer* (New York: Simon & Schuster, 2000), 73–76.

3   http://news.berkeley.edu/2015/01/22/scientists-set-quantum-speed-limit/. The new proof could even affect recent estimates of the computational power of the universe, which rely on the energy-time uncertainty principle.

4   For details, see http://physics.nist.gov/cuu/Units/second.html. The technicalities that surround defining a second are details that do not affect arguments in this book.

5   J. W. Dunne, *An Experiment with Time* (London: Faber & Faber, 1927).

CHAPTER 2

Building a Multiverse

I have deep faith that the principle of the universe will be beautiful and simple.

—Albert Einstein (1879–1955)

The Construction of a Prototype Model
The Proposition and Some Definitions

My motivation and indeed the challenge here was to establish a proper concept of time. But as it turned out, I could not do that until a multiverse was built and the run button pushed. My initial motivation behind this project was to explain at least some of the basis of the now well-documented human experience of being able to see—most often in dreams—what turns out to be our future. It also includes past events in which noting details and verifying them are of great importance. Dreams in which future events are seen and later verified as they actualize in this universe can be quite a profound experience. The long-established term used for this phenomenon is *precognition*, which in the minimal case is a single image. In the maximal sense, it is equivalent to a video display. Its close relative, *premonition*, by definition does not have the visual display. Such displays often contain variable amounts of symbolism such as in Pharaoh's dream in Genesis 41.

It has been said that precognition is the easiest of the paranormal phenomena to document but the hardest to explain. My assessment is that it is not always the easiest phenomenon to document because of the criteria set up by the agencies that oversee and judge the acceptability of each individual case of precognition. You may judge for yourself based on this key chapter that in principle, it is not the hardest to explain at least in the broad aspect. It just took some considerable time to put the explanation together and harder to compose captions for many of the diagrams, which are snapshots of a dynamic system.

The hardest part was to demonstrate that there are two distinct types of time, the foremost one being manifested as a multiverse timing system called type-1 time. It was far from trivial to recognize that this logically led to postulating the existence of a Cosmic quantum computer external to the universes except possibly the Template, the leading universe in the multiverse. Recall that I use lowercase *cosmic* to qualify what is contained or is happening in each universe. This includes the changing outer boundary of each universe. Events in this local system are described in terms of type-2 time, or formal clock time, contained in every universe.

Two Alternate Thought Pathways

There are two ways to start when trying to unveil the nature of a multiverse. The first one is to use the phenomenon of precognition as an indication that there is a place or places not of this universe in which scenes look similar to what they look like here and that people familiar to you are in the scene. This may be a frame grab or a short video-like clip. The scenes may be in monochrome or color, usually vivid. Next follows a rather long-winded mind experiment that is probably trivial for professional philosophers.

A significant observation is that a dreamer can experience a significant lag between when a dream event is entered in a dream logbook and when a matching actualization occurs. If one assumes that the speed of the dream image transmission is constant, this suggests that dreams are coming from different locations; that is, the image path length varies. The simplest case to work with in terms of geometry is a serial multiverse. This further suggests that there exists a copy scheme in which copying runs in the direction from the future to the present as experienced in any one universe except the Template. There is a complication to this picture in a cluster of universes behind the Template because of the development of first precognitions, which cannot occur in the Template. This will become quite evident in later chapters. The implication of this picture is that each universe exists in a time zone different from the universes on either side of it. In each universe the time base is exactly our own familiar clock time, except that there is a fixed time shift between each universe.

The second way to build a multiverse is to search in the field of physics for clues that might appeal more to twenty-first-century physicists. My obvious starting point was the perennially fuzzy but intriguing concept of block time. This has sprung rather dubiously out of special relativity physics; at least it is found in books on the subject. I will just start with it as a concept to be objectively interpreted. It can be taken generally to imply that all Now moments that ever existed are all in existence simultaneously. The single term *time* is now temporarily out of service.

I take a further step by specifying that all Nows are unique and real and thus require real space to exist in. This is provided by three-dimensional spaces each one like space in our universe. This gives a huge number of three-dimensional spaces in a block containing an equal number of Nows. These are the event snapshots.

So far in this thought experiment, thinking is being done in an imagined static state. There is no explicit time in there, only frozen events. Now moments are essentially represented by slices of events; these are analogous to images on a movie filmstrip. The trick is to finally get this unnatural state of stasis moving in some reference framework. Then and only then can time be manifest. That is essentially what I do to arrive at a multiverse. In this mode, I make the phenomenon of precognition (and premonition) a prediction or consequence of the particular multiverse model I build.

The Plan of Action

I generally avoid being held up in this chapter by speculating on mechanisms because they are in the domain of metaphysics. In chapter 5, I attempt to address the nature of the Now. Otherwise, I present someone else's discoveries I consider to have some merit. In fact, in most cases, mechanisms are not considered essential at this level of presentation of the model. A well-known analogy is with gravity in Newton's formulation. He derived a formula for the attractive force existing between two masses in terms of their actual masses, the distance between them, and the gravitational constant G. It works very well, but he did not know the mechanism involved. He also said that time flows (suggesting a mechanism). I and many others argue that it doesn't do that. One just uses type-2 time in his equations, and they work very well for most purposes.

Apart from one elementary conversion expression, there are no equations in the text. I maintain that the job can be accomplished using plane geometry and logical statements just as Newton did for many of his publications before he was urged to publish his coveted principles of the calculus he had developed. In the one case where I do use an equation, it is hidden. That equation involves the generation of the bi-tapered torus that contains and constrains the movement of the universes. The details are not crucial to the large-scale interpretation of the model. The link to the equation revealing these details are in the end notes.

Getting Started: Some Fundamentals—
The Use of Dots, Never Points

In preparing for the construction of the upcoming diagrams, I will explain the essential representations used: lines and dots. A line serves as marking a temporary scaffold for connecting dots (•), which have dimensions. A dot labeled with an N represents a static Now moment in clock time; it could be called a snapshot depending on the context. This is a frozen slice of action that can be studied in relation to other

Now dots. If your digital camera has its time stamp turned on, you will see a red time and date label. That represents a scalar reference time value; in this case, data. Likewise, in the dot, time is in stasis. Next, we look into the term *block time*. This term is a stepping stone in an evolution that leads to the term *block Now moments*.

Assumptions

The fewer the assumptions a theory contains, the better its chances of being accepted. The fact that we are getting set up to unveil a multiverse was a result of making reasonable statements about the existence of an evolving concept (block time). The origin of this term is assumed to have come from physicists, philosophers, or metaphysicists. To give it animation, the term *growing block time* has been used. The block prefix was applied to our universe. This is clearly impossible as it must be extended to involve a multiverse. There are no assumptions involved in the establishment of the first set of diagrams. Beyond that, two assumptions are made. I elect to follow in outline a specific solution due to A. Friedman,[1] who used Einstein's simplified field equations to derive three main types of dynamic universe. I arbitrarily specify the closed (loop) universe, which starts with a genesis event (GE) and closes with an event that would correspond to what has been loosely termed the big crunch. I use the term *Crunch* (for Cosmic crunch) to avoid endorsing a particular interpretation. This will take on more meaning when we examine the enigmatic term *bouncing universe*.

To avoid the mathematical singularities in Friedman's equation for the variation of the scale factor versus cosmic time (i.e., time inside a universe) and because of the absence anywhere of accelerating expansion of that model universe, I have used a substitution: a well-known and well- used mathematical function called the Gaussian bell curve to empirically remove the mathematical deficiencies of the model universe because this would also affect the model multiverse. In doing this, there is nothing that will affect the general results and conclusions arrived at because in the verification stages of the multiverse; only short arcs of the full loop far from the ends need be used.

To remove the result of a mathematical singularity that occurs only in an analytical solution and nowhere else, there is a finite space existing at the start (the GE) and at the end (Crunch) of the loop. With this configuration, there is a finite number of universes in the multiverse, which you will see in chapter 6 agrees with another model (MIW) that appeared while the manuscript was being first externally edited in 2014.

Construction Procedure

Four main diagrams cover a series of elementary steps. No essential trigonometric equations are used in this process though a few mathematical terms and symbols are employed serving in a shorthand role. The tools involved are the most basic forms of logic, elementary geometry, and some reasonable expectations about the nature of extra large-scale Cosmology. The capitalized term is used for the whole multiverse; the lowercase applies to individual universes.

In figure 2.1, diagram *a* shows a linear concept to order a large set of finite static Now moments; this is the simplest configuration possible. The line is arbitrarily straight, but it could equally well apply to a curve, which it will later become. Consider one such Now (denoted N) at some position along the line that serves as a scaffold holding other (invisible) N dots. To make it quite clear that there is no explicit functional time during stasis, the order of events (or slices of them) is given prominence. Thus, the traditional time arrow[2] has been eliminated. The N dots individually represent just part of an event by analogy with the individual frames of a laid-out movie filmstrip, which conveniently provides a graphic display of the principle of cause and effect.

Overall, there is a direction of evolution of a large sequence of events. The relative age of any event snapshot increases to the left. The reason for the origin being to the right is intentional and was determined by previously examining the nature of the graphics that follow.

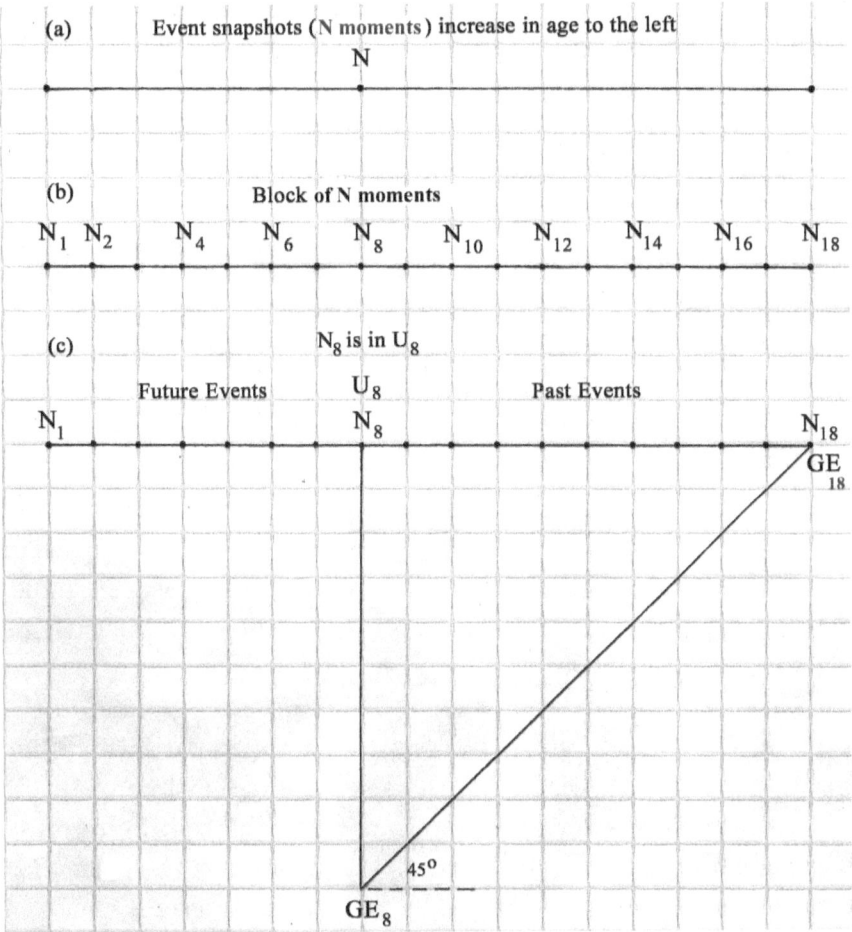

Fig 2.1 First three stages in the multiverse construction. (a) This shows a 'scaffold line' that holds a finite but huge number of 'Now' moments, of which one, **N**, is arbitrarily selected. It coincides with N_8 in panel (b) which shows a few labelled **N**'s where the most recent 'Now' is N_{18}. If each 'Now' is to occur simultaneously, it has to do so in a separate three dimensional space and we are forced to specify that this space is a discrete universe. In (c) N_8 occurs in U_8. All other N_i's occur in universes denoted U_i. The last apparent instantaneous *start location* on the dot-studded 'line' N_1 to N_{18}, is N_{18}, which is where the most recent Genesis or 'Generation' Event, GE_{18} occurs. An animation would show the *line* $N_1 {\to} N_{18}$ moving up the page. N_{19} (not shown) would appear to the right of N_{18}, off the diagram but in line with $N_1 {\to} N_{18}$. Time is suppressed and focus is on *events*. Denoting N_8 as our universe, it can be stated with some certainty that the duration N_8 —N_{18} (as of this instant) represents ~ 13.7 Billion years. Next we take advantage of there being another linear dimension available to represent the past history of our universe. This is represented by the perpendicular line $GE_8 \to N_8$. This line should be the same

length as $N_8 - N_{18}$ in order to preserve the same (time) scale between active ($N_{18} - N_8$) and historical ($GE_8 \rightarrow N_8$) durations. It will be shown in the next figure how these apparent 'distances' are *time-like*. Refer to text for more explanation.

The following expands on what is explained briefly in the caption for figure 2.1a. In diagram *b*, the scaffold line of diagram *a* is populated by a selection of Now moments held in stasis. They show what we may now provisionally call a block sequence of event slices or snapshots that are equivalent to a Now moment in type-1 time. The eternality concept that all moments in a separate event space occur simultaneously is captured here. The thinking is Newtonian. The main element in the thought experiment lies in recognizing that all the N_is (i = 1 ... 18) belong inside an equal number of universes U_i (i = 1 ... 18). This represents in reality a system of dynamic serial universes that have been temporarily rendered static. They evolved in serial order originating from a location at the right end of the scaffold, which I have selected to be where N_{18} just occurred at the extreme right hand dot.

Diagram *c* was developed from diagram *b* as explained in the caption. A small, upward movement of the horizontal scaffold line needs to be invoked to explain what happens next. It is simply a repeat of what happened before to get to this moment. This continuing mind experiment is equivalent to an animation. Each time the horizontal line moves up, it is accompanied by an advance in type-1 time. An instant after N_{18} occurs, U_{18} is officially in existence. It begins to travel up the grid line on which it is situated and moves away from the 45° diagonal line to allow for the appearance of N_{19} and U_{19} (not shown).

To understand future diagrams designed to discuss the multiverse in motion (as in chapter 5), it must be understood that all the N_i's when the system is put into the run mode need a new subfix $N_{i,j}$ in which i designates the universe number and never changes whereas the index j changes continually as the Now's arriving from the left change on a counter which registers the snapshot number.

When our universe formed just after GE_8 (now below U_8), the horizontal line passed through it. The clock for the ensuing universe (U_8)

activated from zero at the instant that GE_8 occurred. The line $GE_8 \rightarrow N_8$ (shorthand for $N_{8,j}$) represents the full history of our universe that has been famously calculated as spanning approximately 13.7 billion years (in type-2 time) using a method made possible from the observations of pioneer astrophysicists V. Slipher and E. Hubble, whose data has been improved upon in modern times.

In diagram c, we can immediately see the existence of two pasts; one is represented by the vertical line $GE_8 \rightarrow N_8$ (which contains a record of our dead past) and the other represented by the horizontal line $N_{18} \rightarrow N_8$ (which represents our live past when stasis is broken). It is an illusion that locations GE_8 and GE_{18} are physically separated; they are separated in duration only in type-1 time as I will elaborate on.

No one knows how many universes are in our multiverse, only that their number is extremely large.[3] Within the multiverse being described, the number is not important to our present inquiry. The eighteen universes used in the diagrams are adequate to describe the multiverse's functioning and provide some answers to several enigmas. In the animation,[4] more universes have been used. I speculate in chapter 5 that the seemingly excessive vastness of a universe serves a need to protect itself and that the even more excessive size of a multiverse serves another purpose connected with our existence.

Few people have ever considered more than one type of time. I have introduced type-1 time, which is actually a timing system in place before our GE. Whitehead's formal time (my type-2 time) came into existence with the invention of the clock. Recognizing these two times as fundamentally different yet occurring together for different purposes is essential for understanding figure 2.1A, the Blueprint. It represents a completed development of figure 2.1c.

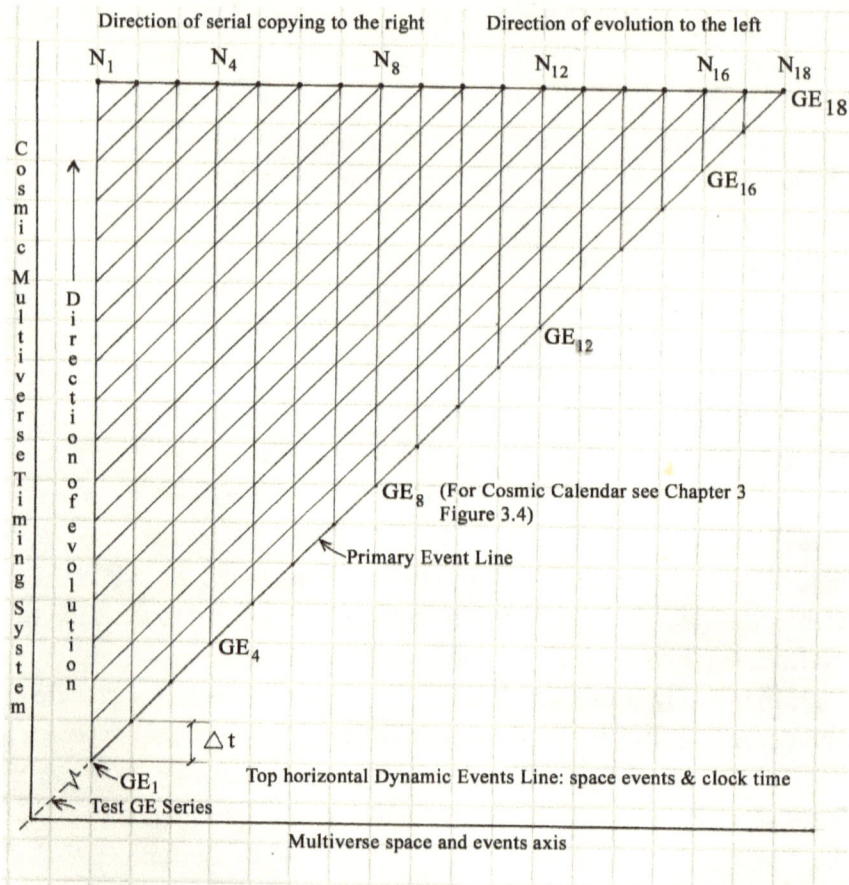

Fig 2.1A Diagram was developed from Figure 2.1(c) and is called *the* **Cosmic Blueprint**. The lowest diagonal line is the locus of all the Genesis/Generation Events. All the other diagonal lines are (past) Event Lines which connect the same (selected) event (or a part of it) that occurs in each universe. Inside the triangle the vertical lines depict *trajectories of evolution* for each universe and all its contents. The (inside left) vertical axis "**Direction of evolution**" refers to all those lines. The horizontal matching axis: "**Top horizontal Dynamic Events Line: space, events and clock time**", refers to the upper dot studded line, where all the action is when stasis is replaced by dynamics. Every line *below* the upper row of dots is history. N_8 occurs in U_8 which is our designated universe. The "Test GE series" is explained in the text. The *last* part of the construction involved adding in the 'absolute' reference frame axes which are *outside* the multiverse. The *outside* vertical axis "**Cosmic Multiverse Timing System**" references the other type of time ('Type 1 time'). This involves a time step Δt (marked at the lower corner of the triangle); it is a manifestation of a stable frequency ($1/\Delta t$) –see text. The *outer horizontal* axis labelled "**Multiverse space and events**" represents processes involving the whole multiverse.

I will consider several aspects of the Blueprint next. The Test GE Series label is pure speculation in that it was inferred to have occurred after asking, "What preceded GE_1?" The only rational answer is that an experiment was being run to determine the values of certain physical constants that would generate the right conditions for life to develop in a universe. So why begin a copy procedure? How was the value of Δt determined? Some short responses follow. The first question will be answered later.

As far as the ultrashort duration Δt is involved in representing another time, it is more fundamentally considered in terms of a frequency, which is simply $1/\Delta t$ expressed in cycles per second or whatever the units were for the engineer who designed the multiverse. Thus, the system may be compared to a digital watch employing a quartz crystal oscillating at its fundamental frequency and driving a counter and divider that produces time pips at one-second intervals. In the multiverse, the frequency f controls the rate at which the multiverse evolves through a Cosmic quantum computer.

The Cosmic Blueprint diagram represents a dynamic system that is presented here in a semi-symbolic, not explicitly physical form except for the row of primitive dot universes that constitute the dynamic events line (*DEL*), the only moving part, which progressively moves upward. The two types of time are combined within it. Extraction of physical information from the Blueprint is possible.

First, the event lines and the evolutionary lines indicate that events are being copied from the lead universe U_1, called the Template (containing $N_{1,j}$) to the universe behind it and so on along the series. The $N_{1,j}$ values are generated in the Template according to the laws that we are familiar with. Time lags are automatically built in to these succeeding universes right back to the GE generation.

Second, the upper row of universe dots now labeled along to N_{18} grew from a first position (following a duration Δt after the GE_1 event) when there was only N_1 and N_2. The actual Now moment N_1 ... etc. is simply a cover label for a stream of snapshot files containing a slice of all the events in that particular universe. The length of this horizontal line of dots progressively increased (from the right) as it moved up vertically, regularly picking up a new universe as it went. But this movement is a progression in type-1 time at a fixed location as you will soon see in the next figure.

The two direction statements at the top of figure 2.1A have a connection to Δt, the fixed duration between GEs, the duration between two consecutive positions for all the universes, and the duration between copy commands that transfer information files from one universe to the next. Thus, the information transfer is traveling to the right along the horizontal row of dots (the *DEL*). On the other hand, the continual addition of new universes at the right end of the line means in reality that all the dots are moving left. The only physical circumstance in which this could happen is if the GE's are all occurring at the same location. Thereby, the Primary Event Line depicts a process taking place in time only. The next diagram is spatially oriented and time (consistent with the Blueprint) has to be manually labelled on it.

I opted for the closed-loop interpretation of each universe being guided by the principle of Occam's razor, the principle of efficiency, Einstein's belief in the ultimate simplicity of the cosmos, the MIW approach,[5] and arguably, support for intelligent design. Thus, figure 2.1B emerges from the Blueprint and is called the Cosmological model. In verifying the model in later chapters, figures 2.1A and 2.1B are used as is appropriate to the application.

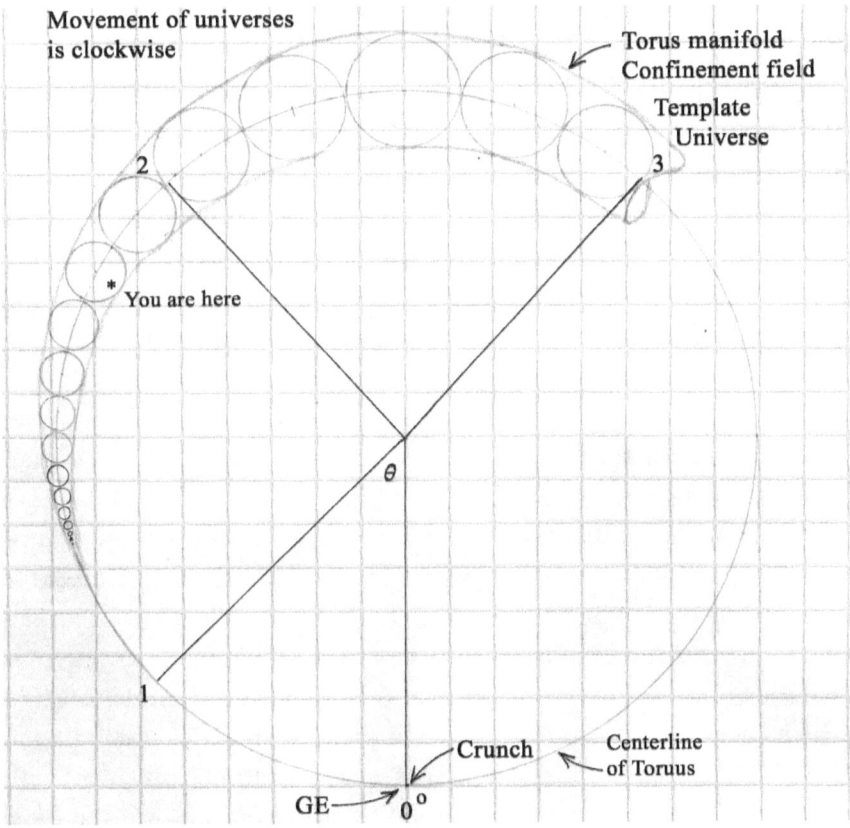

Fig 2.1B Cosmological model, derived from the Blueprint. All universes are generated at the same location (GE), and move clockwise from there. The circumference of the circle on which GE lies serves as the centerline for the 3-d torus manifold that is symmetrical and filled with the 'quantum field'. It pervades the universes. The 'vertical' line connecting the center of the circle to the GE location serves as defining where the angle $\theta = 0°$. The location **1** corresponds to $\theta = 45°$. Universes exist from $\theta = 0°$ to 72° but at this scale are 'invisible'. Site **2** ($\theta = 137°$) is a curve inflexion location where the torus expansion changes from accelerating to decelerating. At $\theta = 180°$ the maximum size of the universe occurs. The reverse inflexion location occurs at site **3** ($\theta = 223°$). As universes move around the loop they *continually* changing their diameter and receive their next Now moment from the universe ahead of it, except the Template, which generates the original events that are continually copied to its immediately following universe. Thus the copy scheme operates counter clockwise. At all *locations* (**1, 2 & 3**) the *same event* always occurs as the universes move through them. I have arbitrarily stopped the Template at the **3** position where it is in the process of switching from accelerating contraction to decelerating contraction on its way to the Crunch. The * labelled 'You are here' refers to a hypothetical position of our universe in accelerating expansion. The atmosphere of metaphysics that this diagram evokes will be redeemed by the volume of empirical evidence that it supports.

In figure 2.1B, universes are made to move clockwise through the torus, only part of which is shown. In the first half of the circuit, this is the reverse direction as shown in the Blueprint, but this was a result of making necessary decisions on directions of movements needed to be adopted in the Blueprint. Now you can easily see that each universe is generated at a single location at intervals of Δt. The shape of the containment field (inside the torus) has been engineered in a sense to conform to the geometry of the Gaussian bell curve spun around the straight line of figure 2.1A that will become the circumference of the large circle. This gives the torus three dimensions, so its curved surface defines a mathematical manifold. Each universe thus goes through a size change in accordance with this manifold into which it must fit exactly.

The horizontal line of dots in figure 2.1A is now replaced by universes according to the size changes as required by the Gaussian curve formula which is a replacement of the cycloid curve that Friedman obtained. This is metaphysics by necessity. But the important point is that the multiverse displays an evolutionary character mimicking that which each universe goes through. In the new empirical formulation there is always a finite size of matter at the GE location (coincident with the Crunch) due to an arbitrary splicing of the ends of the synthetic Gaussian torus. Now this is an engineering problem not a physics one but it certainly is an effective way to get rid of the two mathematical singularities which occur in the Friedman solution. Einstein did not seem interested in this situation except that it might be reflected in one of his quotes paraphrased as: In so far as the mathematics seems to be correct it does not guarantee that it corresponds to reality.

The event lines in the Blueprint have their expression as fixed locations in the Cosmological model. At any selected location around the large circle's circumference is a location where the same event always recurs in universes that pass through it on their way forward (i.e. clockwise). This is meant by the notations posted at the top of the Blueprint diagram. The universe containing the label: "You are here" is U_8 in the Blueprint. It is placed there because current estimates indicate (from astrophysical data) that our universe's expansion is accelerating. Location 2 between the next two universes is where in this model all universes change from accelerating expansion to decelerating expansion.

Some cosmologists speculate that this acceleration indicates an open Friedman-Einstein universe solution in which the universes will expand forever. This is a very dismal prospect in the long run as it implies that life would be slowly decimated forever. It would be hard to imagine why the architects of the multiverse would select a solution with such an outcome. I would want to see more data. Though the theoretical closed-universe solution doesn't correspond to recent empirical data, this more than anything else suggests that refinements in the theoretical equations of general relativity are needed or how simplifications of them are handled.

The Gaussian equation contains an exponential term, and that enables a finite-sized particle of matter to exist at the GE /Crunch location after one complete circuit. Nevertheless, such a condition doesn't affect the arguments I put forward in this book because I have selected locations far from the GE/Crunch location. Furthermore, this is an important condition when we come to disqualifying the idea of bouncing universes in the fifth essay in chapter 6.

An interesting methodological observation is noted with the building of the Blueprint. It is normal practice to start a geometrical exercise by drawing reference axes. This could not be done here because I was working in the mode equivalent to reverse engineering. The final form of the Blueprint was not anticipated in the early stages. The last step, of adding two pairs of orthogonal axes, thus tying the diagram together, brought closure to the exercise. The two types of time are implicitly contained in the one diagram. By comparison, the Cosmological diagram is basically spatial and can have time (the 3 + 1 neo-Newtonian formulation) easily superimposed on it.

An Interim Summary of the Use of Cosmic Multiverse Time and Cosmic Time

It is useful to reiterate some qualifier terms that contain the word *time*. I haven't encountered any reason to become involved with the space-time concept of the theory of general relativity. Regarding the rules that apply outside the individual universes, the noted American-British physicist and philosopher David Bohm said that the physics we know about inside this universe doesn't necessarily apply beyond the confines of this universe, and that must now apply to all universes. In figure 2.1A, the term *Cosmic multiverse timing system* is used to refer to a time concept that exits outside all universes, though in each universe, it has an imprint that will become clear when dealing with the hypothesized mechanics of the Now moment in chapter 5. It gives rise to the term *type-1 time*. Our understanding of it should be in the form of a frequency that is fundamental to the operation of the multiverse as a whole as discussed in chapters 5 and 6.

The term *cosmic time* is used inside our universe (and hence all others) as being equivalent to clock time: I call this *type-2 time*. Note that the cosmological equations that resulted from the solution of Einstein's simplified general relativity equations gave a cosmology that doesn't possess a relativistic space-time character. Rather, the scheme is Newtonian. Newton's universal gravitational constant G appears in the equations. Indeed, one still finds in the physics literature statements such as, "The current model of cosmology permits the existence of an absolute background time."

The Friedman graph for the evolution of a model universe is expressed using orthogonal axes: the vertical (up-page) axis is the scale factor (taken as being proportional to the instantaneous radius of the universe), and the abscissa (cross-page axis) is labeled cosmic time, which is consistent with this chapter.

Explanations of certain concepts can sometimes be accomplished using the Cosmological model and sometimes using the Blueprint. An example where the Blueprint is used to examine the concepts of past, present, and future from two viewpoints is presented next.

In figure 2.2a, the diagram shows how a non-precognitive person views tensed time. Figure 2.2b shows what a precognitive and retrocognitive person would see under the right conditions. The dynamic events line (*DEL*) is taken from figure 2.1A.

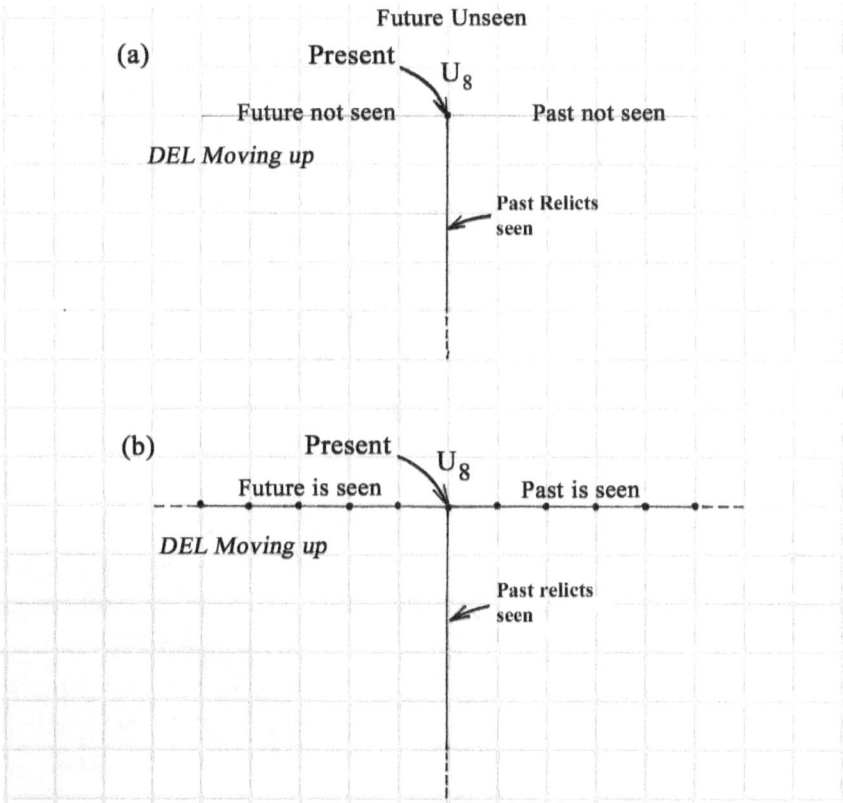

Figure 2.2 Diagram (**a**) is the case for people who cannot see into other universes and only relicts of the past are to be seen everywhere. The grand illusion is that we reached the 'Present' in this universe (U_8) by what happened in the *seen, remembered* and *recorded* past. The 'Future Unseen' label–while you are at the 'Present' applies to both (a) and (b). In case (**b**) where people can see into other universes the actual situation is quite different as you have seen in the Blueprint. In animation the *Dynamic Events Line* (*DEL*) moves up. We exist at a T-junction, where some persons will be able to see not only the visible passive past, but they can also 'connect' with both the *active* past and the *active* future, which according to the Blueprint is occurring simultaneously with the present–*right* 'Now'. Such a view supports eternality and seems to be in agreement with what Einstein and other physicists have said —from which the concept of block time was proffered. This also comes from many cultures spanning many centuries. The serial time zoned multiverse under stasis, as in the preceding figures, embodies this concept.

The Template Universe U_1

I will fully demonstrate the significance of the Template in chapter 5, in which several new concepts are introduced. Here, it is sufficient to state that all our physical lives originated in it and are copied in the reverse direction to the forward movement of all universes shown in the Cosmological model. This is where the question of the purpose of this multiverse can be addressed. The key to understanding that question is to know that we have a companion spiritual energy body that did not derive from this physical multiverse. Any person who has had an out of body experience must be aware of this fact.

Figure 2.1B raises some questions. What happens when the multiverse fills up? If left to recycle, the multiverse will become an exact repeat of the first cycle though it will contain remnant universes of the first cycle until second closure. It has already been determined that the fundamental universal constants have been finely tuned for producing an environment conducive to the development of organic life. There is no use in changing the constants in this type of multiverse because as soon as the first new universe was generated, the future to present COPI procedure would automatically overprint the effects of the new command given it. To generate new content into a closed multiverse, a completely new torus and its energy field would have to be built.

I am aware that there are guidelines—in effect established requirements—inherent in the presentation of a new scientific theory which is being done in this book.

1. Your theory must include a detailed mechanism that can be described by known physics. This cannot be satisfactorily done here by me, but there are several persons who could contribute to describing the Cosmic quantum computer — given this type of serial multiverse.
2. Another law says your theory must make predictions or at least explain long-standing enigmas. This will be done in the following chapters.

3. There is a questionable law stating that your theory must be falsifiable. I am not certain if the STZM theory is falsifiable, but if the theory stands up to requirement number 2 —which I believe it does —this should provide an exemption from law number 3.

One amusing conjecture is that, "If ever anybody discovers exactly what the universe is for and why it is here, it will instantly disappear and be replaced by something even more bizarre."[6]

If so far, the STZM model is thought not to qualify as a true scientific theory it could be classified as a metaphysical theory. Philosophers do that regularly. Some physicists who have respectable positions in mainstream physics exhibit another side of themselves; they indulge in metaphysics, and a few are even known to write science fiction novels. They will even write one- to three-page chapters in books edited by John Brockman, for example: "What we believe but cannot prove." This is how I learned about psychologist Susan Blackmore,[7] who believes she does not have free will, is unable to prove it, but is comfortable with living in that condition. The STZM model demonstrates that her belief does have a basis.

In the next chapter, I will provide the empirical evidence that except in the Template, we do not have free will —with the special case that this may be overridden in universes situated closely behind the Template. This is demonstrated in chapter 5.

In the last reference, Lawrence Krauss importantly wrote, "I believe that our universe is not unique,"[8] and he gives plausible reasons for holding that view. Not only have I demonstrated that this assertion is in accordance with the STZM model, but I will also provide a new and different view on what Krauss further writes.

> The existence of many different, causally disconnected universes—regions that we will never be able to communicate directly with and thus forever out of reach of direct empirical verification—may have significant impact on our understanding of our own universe.

Krauss's case has its origin in a particular interpretation of quantum mechanics implying the existence of far-different universes. This is an unverified hypothesis that is incompatible with the STZM.

In what I have written so far and in the next chapters, I demonstrate that all the universes in the STZM model *are causally connected*, and we (but not all of us) communicate with other universes, or other universes communicate with us, on a fairly regular basis; *interconnectivity does exist*. This is backed up by what I have written in essay 4 in chapter 6, in which I summarize the many interacting worlds model that has strikingly similar properties to the multiverse model presented in figure 2.1B.

Notes

1   A. Friedman. His important contribution was published in Germany in 1922. Solutions of Einstein's equations and the graphs for universe size changes with cosmic time may be found in www.uncw.edu/phy/documents/TheFriedmannEquations.pdf. The graphs are only qualitative.

2   The term *arrow of time* has no place in this model. A raw time value (or a time difference) is a |scalar| quantity, not a vector. Clocks were invented to expressly produce linear clock-time data, which should be considered only as data —real numbers. Use of time data enables the characteristics of movement (such as accelerating, constant, and decelerating motion) to be determined. There is nothing magical about it. The direction of motion of objects is what creates the vector concept and hence, arrow of motion should be used. Readers who are versed in practical mechanics already know this.

3   An early calculation derived from string theory gave a value of about 10^{500}, and this has changed as newer models with more dimensions have appeared. One model suggests a much higher value of 10 to the power of 10 raised to the higher power 16! That is: $10^{10000000000000000}$ proposed by A. Linde and V. Vanchurin as reported in *New Scientist* in 2009. This value represents a cut down from an even more staggering number because it is beyond the absorption power of the human brain! That reduction is a result of an anthropocentric view of the matter. These values don't apply to the STZM model because it isn't based on string theory or any other multiverse theory.

4 The animation, not included with this book, works with eighteen universes for the Blueprint and about a hundred for the Cosmological model. It will be made available after updates of v. 1 are made.

5 In chapter 6 essay number 4 is a description of the MIW approach.

6 D. Adams, *The Hitchhiker's Guide to the Galaxy* (New York: Crown, 1979). He manages to avoid any mention of responsible entities behind the scheme. This would require a new set of fundamental physical constants and compute-and-paste operations to do what was threatened.

7 S. Blackmore, "It is possible to live happily and morally without believing in free will" in J. Brockman, ed., *What we believe but cannot prove* (New York: Harper Perennial, 2006), 40–41.

8 L. M. Krauss, "I believe that our universe is not unique," in J. Brockman, ed., *What we believe but cannot prove* (New York: Harper Perennial, 2006), 214–15.

CHAPTER 3

Hacking into the Multiverse

Overview

The main purpose of this chapter is to find a property of the STZM model that can cast light on the occurrence of the phenomenon known as precognition and its first cousin, premonition. In so doing, this exercise provides a start to the verification of the model. Recall that I arrived at the structure of this new multiverse scheme by using a combination of existing concepts in physics, metaphysics, geometry, and logic. I found that the multiverse model was amenable to qualitative testing using other well-known phenomena. I deal with a few additional topics: déjà vu (which might share an affinity with precognition but in a latent form), (re)incarnation, three challenges that can be substantially addressed, and the well-known grandfather paradox.

I took more topics from this chapter that fit into this chapter's aims when it exceeded a critical size and relocated them in chapter 5, which then was allowed to exceed a reasonable size because of discussions as new material appeared on the Internet. These three challenges will be of interest to many philosophers of science.

Despite a rapidly growing realization that there might be other universes beyond ours, there has not been until now any light thrown on the question of how our universe relates geometrically to these other universes. In a sense, this chapter offers an introduction to the capabilities of the STZM model in terms of solving long-standing issues that have challenged human minds over centuries and even millennia. I felt it necessary to first verify the STZM model using just a few examples in a short chapter. Already-convinced readers may proceed to chapter 6 if they want a succinct summary of the skills of the model. Those who want more-convincing data can tackle lengthy chapter 5. For those who need a background survey of time-haunted people, chapter 4 offers an account of some of the main characters in the twentieth-century struggle to understand the meaning of time and its tenses. Many of the characters who are surveyed appear strongly interconnected but not always positively.

The Process of Verification of the Model

There is a significant observation worth mentioning here. It has to do with scale changes that need to be traversed while discussing a phenomenon, precognition being a prime example. Here, the location of the percipient is compared with the location of the target site from which the information comes: (1) another universe —if locked onto it or (2) a group of universes if speed scanning is being done. This is pure conjecture. Realize that I am trying to use humanly experienced phenomena (dreams, say) in just one universe to verify the huge multiverse model. But this view needs to take into account the fact that even these latter so-called local phenomena actually depend on the existence of the multiverse. Our modern thinking seems to have adapted to adjusting quickly to immense scale changes as shown by several Internet videos of scale changes from an electron to galaxy clusters.

Various phenomena usually labeled *paranormal* are experienced by a significant percentage of the human population regardless of age, gender, class, religion, or education and throughout history. There is a new word for this list of phenomena: *preternatural* which I endorse. In stark contrast to those who tend to strategically conceal their ability in this area, a few people choose to discuss these experiences openly. My view is that precognition, retrocognition, premonitions, and déjà vu should not be called paranormal along with such weird phenomena including levitation, spoon bending, poltergeist activity, and perhaps others as may be found in most encyclopedias that deal with unexplained occurrences. The phenomenon I attempt to address here clearly involves information transfer from what can be seen as past and future locations along the dynamic events line in the Blueprint or on the loop trajectory of the Cosmological model. As such, these phenomena form a valuable source of useful data for verifying the existence of a multiverse. In this chapter, it is the Blueprint diagram (figure 2.1A), that gets the most use.

Whereas the topic of religions is outside the scope of this book, I do use one Old Testament text that transcends religions; it addresses time in such a way that the author, arguably Solomon, seems to have had an implicit grasp of the existence of other worlds. This is found in the philosophically yet also practically oriented book of Ecclesiastes. That book is an edited copy from Kohelet found in the Hebrew Tanakh. Though this is only one of two other possible interpretations, the analysis of the extracted verse from both books forms the first of the three challenges. The second challenge is based on T. S. Eliot's description of time found in "Burnt Norton" in the "Four Quartets," and the third challenge is one formulated by a respected contemporary philosopher Dean Zimmerman.

To finish the chapter, I provide an analysis of the well-known, so-called grandfather paradox and my dissolution of it. This makes use of the Cosmological model though there is no need to provide a special diagram; reference to figure 2.1B is sufficient.

Phenomena Seen to Be Explained by the STZM Model

Precognition and Retrocognition

Many people who experience precognitions retain the impression that these events are typically spontaneous but strangely sporadic and hence that they do not willfully control them.[1] A minority seem to be able to control the occurrences of these preternatural phenomena under special conditions. Despite the fact that this phenomenon has been studied intensely for well over a century by many international groups that have included prominent physicists, the specific mechanism involved has not been revealed. This is because the physics involved is outside the scope of current science. Consider this statement.

> We have come to believe that the here and now defines the limits of our existence. The past, we say, has already happened and no longer exists; the future has yet to unfold. The present is all there is. To this, premonitions say, "Wake up. The evidence for a larger world is staring you in the face."[2]

These are the pressing words of Larry Dossey, MD. He uses the term *premonitions* to include precognitions, whereas I as do others use these related terms as standing in their own right. In this book, precognition connotes a visual graphic of a future event whereas premonitions are commonly associated with feelings and sometimes a heavily symbolic snapshot that may suggest a warning about an upcoming event.

Concerning dreams, Dunne told us in the first part of *An Experiment with Time* that according to his assessment, on average, any given dream that is significant in the context of a future event is composed of images from the future mixed with images from the past in the approximate ratio of 1:1. However, symbolism is a complicating factor that may be involved. This concept is found in the writings of Carl Jung, whom we meet in chapter 4.

Precognitions have a long history and are recognized to occur in the Bible.[3] Concerning precognition, Dean,[4] for example, stated that this "aspect of parapsychology" is "the hardest for us to understand yet the easiest to experience and to test."[5] I have read many statements similar to this; indeed, the *go* of it has taxed the human power of reasoning for millennia. The multiverse scheme immediately provides a large-scale explanation for precognition but only in terms of pathways for information transfer. Here, I need to use only the Cosmological model. Clearly, from figure 2.1B, the pathways from our universe (U_8 in figure 2.1A) to other universes all appear to be equally available. This concept also applies at N_i dots occurring along the horizontal scaffold in the Blueprint provided that the N_i dot lies in a band of diagonal event lines beginning at the dawn of human consciousness and ending at the event line where a hypothetical species extinction occurs. Depending on your objective and focus, any given clock time, date, or year can be used by selecting the size of the event of interest. This may be the full spectrum of events in the universe, just one local event, or to the snapshot level equivalent to a frame grab. The ultimate thin slice of clock time takes you down to the duration of the Now moment itself.

Either precognition or retrocognition can occur in dreams or in waking visions. I will offer examples of these later in the chapter and in the overflow relocated to chapter 5. Of the two, precognition is the most studied for two reasons. First, it enters into discussions of free will versus predetermination. This latter topic is the subject of another subsection. Second, it is involved in the apparent paradox of retrocausality, backward causation, which will be dealt with in a later subsection. I next remark briefly on the fairly common phenomenon of déjà vu.[6]

Déjà vu

This is the common sensation of having been there before, heard that before, and having done that, etc., as relayed to our senses. The phenomenon was documented in detail for the first time by a French researcher, and each sensation was tagged with a unique French language identifier. However, the *vu* qualifier is evidently now used loosely to cover all the types. A vast knowledge of this experience is available on the Internet, from books, and from almost every third person you meet, so I will keep this section brief.

Déjà vu occurs as a rather subtle recognition experience without causing any alarm; it is not unexpected as it is lagged or delayed in duration with respect to the actual event that happens in this universe. For example, two people are having a conversation. At one moment, the person experiencing déjà vu exclaims, "Wait a second! This conversation is familiar!" This is the audio version (say déjà audio). An expanded version of déjà vu, where other senses such as smell, taste, or touch are also experienced, has been termed déjà vécu. There are several different theories but no clearly accepted one regarding the possible mechanisms involved.

The STZM model might be used in this case. It is possible that déjà "—" information may use similar pathways as in precognition and premonition within the field that fills the torus containing the multiverse. Déjà "—" could be a case of slightly delayed recall of some forgotten precognitive dream (residing in the subconscious) that suddenly surfaced in the mind right after the actualized event occurred (as if it was acting as a subconscious memory cue). In that respect, déjà "—" has rather fittingly been called future memory.[7] An interesting case of a person who experienced precognition and déjà vu was U.S. Army general George Patton He also claimed to have experienced visualizing a past life. It is not unusual for the same person to experience several psychic phenomena.

Incarnation and Reincarnation

As is the case with prophecy, belief in reincarnation[8] has very ancient roots. Though not commonly discussed, it is nevertheless a well-known phenomenon existing in many cultures and countries over most of the last three millennia. It is apparently not a basic belief of the early Christian Church though several hints of it are claimed to exist in the gospels and elsewhere; this suggests that more-explicit elements may have been removed or that a selection process for books of the New Testament may have eliminated explicit cases of recall of a former life. This is perhaps not surprising as emphasis has been put on the present life of individuals. The protocol involving transmigration of souls as spirits is sometimes included in studies by a limited number of researchers.

Given the existence of eternalism, or eternality, a spirit may in principle choose not to reincarnate into the same universe they lived in before but another one. The multiverse offers a much expanded choice of bodies for spirits to select a physical body than a single universe could offer. This solves the wait listing problem.

Reliable anecdotal accounts suggest that after waiting a particular duration (sometimes as short as less than a year to a few years), spirits may reincarnate into a chosen near-birth body in this universe. By association, the same must happen in other universes. The following is a generalization of what were specific cases. On record are accounts[9] of children (typically between ages two and six) who could recall previous lives. The memories must have been retained by the soul. Some of these reincarnations occurred in the same village that the (now) deceased person once lived. With such a short return time for transmigration, the opportunity exists for the researcher to check what these children have told their parents. During the short time the child's soul was spirit, the villages wouldn't have changed much and there might be people still alive who remembered the deceased person whose soul (preserved in the spirit state) had before birth entered the child in question. The child would often recognize people in old photographs, recall names, and recognize houses and streets in the village. These cases typically involved early death by accident or disease.

Another way to study past lives of a person is to use past-life regression conducted by a psychologist employing hypnosis[10]. Yet another way that a past life may be revealed is by spontaneous awareness through dreams or visions. General Patton had such impressions, and it is possible to access compelling accounts in videos on the Internet that show how dreams of a past life can affect a person's life.[11] One notable inference is that these memories must have arrived with the reincarnating soul.

Skeptics attempt to discredit the practice of investigating past lives as well as classifying precognitive dreaming as an impossibility of the third kind. Some try to debunk organic evolution and many other things outside their ability to comprehend. I have a response to this.

In chapter 2, a new multiverse model was constructed in a typical length for an average journal article. It required verifying in several different areas, which took chapters 3, 5, and 6 to accomplish, although there is some slightly repetitious material in chapter 6. This indicates that an overwhelming amount of material was delivered to "verify" the model and solving puzzles or paradoxes using appropriate parts of the geometrical scheme.

The most important question to keep in mind at this stage is this: "If you agree that a multiverse exists, what do you think its purpose is?" On this score, of the ten other multiverse schemes, usually identified with physicists, I haven't so far encountered any mention of a purpose for a multiverse. Oddly enough, the only person who consistently expounds on a purpose-guided universe and is not averse to the existence of spirits doesn't admit to a multiverse. Is this a case of religious influence or an application of Occam's razor? Why settle for billions of trillions of universes when one will do? The answer is that with only one universe, the question of its origin is left totally unanswered and no one would experience precognition.

The only logical explanation I can arrive at is that the multiverse loaded with multiple copies of individual lives was built for multitudes of spirits to incarnate into so they could experience a physical life — just as Pierre Teilhard de Chardin wrote in the middle of the twentieth century. The reason for the need to know this is currently outside my personal experience base and not of immediate concern for this book. Others[12] in our own culture have temporarily entered the eternal life zone to find out just what was happening over there. But there were even others in different, ancient cultures who discoursed on a meaning of life that included our soul's existence as spirit. That is the holistic and most acceptable view as I now see it.

A Search in Unlikely Places for Substantiation of the Multiverse Model

In a special exercise to follow, I will address some selected challenges to make the capabilities of the multiverse model more convincing and show that other people have had the same serious concerns as I and many others have had about how to deal with time; we are the time-haunted people. In a sense, the challenges presented themselves to me as puzzles that needed to be solved to make any progress in our thinking about time. Much of the confusion has arisen as a result of our general unfamiliarity with thinking outside the universe.

The first challenge is an ambiguous passage from Ecclesiastes, the next one is an excerpt from one of T. S. Eliot's famous poems containing a time theme, and the last one I call Zimmerman's challenge after the philosopher who posited it. They can all be addressed by using a short section of the Blueprint. The Cosmological model may be consulted to bring the discussions into real geometrical focus.

The Three Challenges

Interpretation of a Time Theme in the Book of Ecclesiastes

Having linked the multiverse to matters of a spiritual nature, I must make it clear that I am not deliberately making an association with religions by selecting a passage from the Old Testament. I did this because there was the possibility of associating ancient ideas about time with time as it appears in the multiverse model.

This self-generated challenge resulted from a question I put to a friend: "Dawn, is there anything in our Bible that suggests deep knowledge of time and particularly tensed time?" She immediately replied, "Look in Ecclesiastes 3:15." Back home, I pulled out my King James Version of the Bible and studied those words of wisdom. I also acquired a copy of the Jewish Bible in which the source book is referred to as Kohelet.[13]

While not the shortest book in the Bible, the modern version of Kohelet (K) is quite brief by biblical standards. The name means the preacher or teacher and is considered by biblical scholars to have been written by King Solomon near the end of his life in 931 BC. Consider K 1(16): "I have acquired much wisdom, more than anyone ruling Yerushalayim (Jerusalem) before me."

Noted American novelist Thomas Wolfe wrote about K as follows:

> Of all I have ever seen or learned, that book seems to me the noblest, the wisest, and the most powerful expression of man's life upon this earth—and also the highest flower of poetry, eloquence, and truth. I am not given to dogmatic judgments in the matter of literary creation, but if I had to make one I could say that Ecclesiastes is the greatest single piece of writing I have ever known, and the wisdom expressed in it the most lasting and profound.[14]

Its brevity and the recurring time theme we are interested in here along with other passages that seem to be in the style of a rant suggests that long ago it could have been down sized.

In K 1(9) appears this statement: "What has been is what will be, what has been done, is what will be done, and there is nothing new under the sun." Likewise in K 1(10): "Is there something of which it is said: 'See, this is new?' It existed already in the ages before us." For a one-universe scenario, this is equivalent to saying that history repeats itself, but that is the naïve view. Historical events are never repeated again exactly as they did earlier in the single universe case. Then in seemingly more explicit tones, the verse that Dawn referred me to, K 3(15), states, "That which was—is here already; and that which will be—has already been, but God seeks out what people chase after." Note the repetitiveness of the time theme much as is in the next challenge.

This last quote is now compared with Ecclesiastes 3:15, E 3(15), in the King James Version (1769): "That which hath been is now; and that which is to be hath already been; and God require[s] that which is past."

I will try to explain all three clauses of K and E in the context of the multiverse Blueprint; together, they seem to add to the breadth and depth of the whole verse: the whole is greater than the sum of its (three) parts. Taken literally, this scheme seems to read as if it is identifying with hard determinism. The perplexing part is to figure out how K managed to compose this verse so that it appears to support the STZM.

The use of the determined word *will* in K projects definiteness or determination, whereas in E, that projection has not exactly been preserved. Yet the phrase "there is nothing new under the sun" goes along with the multiverse model. Further, to say, "It [an event] existed already in the ages before us" can be read as pertaining to our future already existing in the STZM universes ahead of us. That is, the event occurred before we experience it as delivered by the COPI procedure.

Nevertheless, K's statements so far can also be interpreted as resulting from believing in one universe with the proviso that the events in different "ages" Solomon was comparing are only similar. For example, though the first and second world wars were massive wars involving similar adversaries and lasting five years, they were technologically very different and involved different people. Consider next that the first two phrases of 3(15) of either K or E seem to suggest more than one universe; otherwise, the second phrase is redundant. It can easily be cast in terms of many universes as I will show. Thus, taking a more literal slant on K, I maintain in principle that the STZM is still in contention to address the challenge.

Now in K, the third phrase differs from the one in E, which is comprehensible in terms of the STZM by exhibiting the past as a simultaneous reality and therefore allowing it to be studied by spirits—regular ones or special ones. As a result of these multiple interpretations, I decided only as an exercise to demonstrate just how versatile the STZM is although its use here could be argued to be on dubious ground. This is particularly so when we consider that the Greek philosopher Anaximander (ca. 610 BC–ca. 546 BC) was credited as being the first to get or toy with the idea of the possible existence of multiple dynamic worlds three to four centuries after Solomon probably wrote Kohelet.

Despite these differences in interpretations, I decided to construct figure 3.1, in which I have reproduced an enlarged cutout of the Blueprint. I demonstrate how E 3(15) can be interpreted in four steps in the geometrical framework of the diagram. As with many of the other diagrams used to verify the multiverse model, there is no need to worry about what is happening near either of the ends of the multiverse loop. Let stasis occur in the dynamic events line *(DEL)*, which is conveniently done for us on paper. The Now labels have not been shown, but they are there as N_i inside each universe, U_i. Here, the $_j$affix on N_i is unnecessary because any arbitrary value would suffice under stasis.

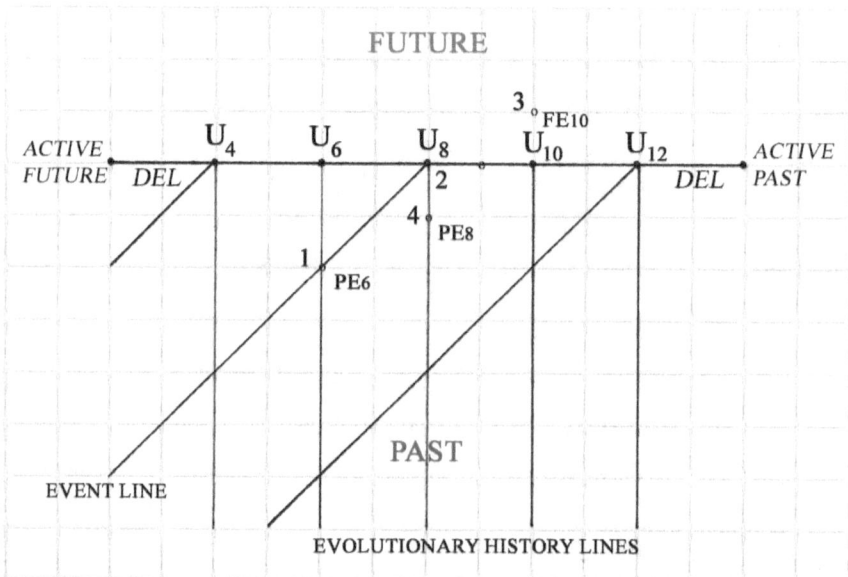

Fig 3.1 Each black dot on the horizontal Dynamic Events Line *(DEL)* represents a universe U_i associated with an $N_{i,j}$ (not shown) where, *for any one universe*, the **i** stays the same but the **j** value is constantly changing in the normal *dynamic* state every Δt. **PE** = a Past Event snapshot; **FE** = a Future Event snapshot. Each snapshot is labelled with a **j** number. For this exercise I am dealing with blocks of event slices within space provided by each universe. Here is *one* version of the sequence of steps (numbered **1** to **4** on the diagram) in interpreting Kohelet 3 (15) abbreviated as **K**:

Step 1: "*That which hath been*" is represented geometrically by **PE**$_6$ (location #1) which is in the past for U_6. It is on the evolutionary line of U_6 at the intersection with the diagonal **event line**.

Step 2: Run diagonally, through millions of billions… of 'Now's', along the **event line** up to *our universe* U_8, which is marked as location #2. This corresponds to where **K** states: "*is now*".

Step 3: To address the second clause, go up to location #3 labelled FE_{10} which lies on an (invisible) extension of the evolutionary line of U_{10}. It represents a *future* **event** for U_{10} which is currently frozen at $N_{10,j}$. This is the event described by: "*that which is to be*" in U_{10}.

Step 4: Go diagonally down the two small open (.'s) to location **4** (a dot) which marks PE_8. This is the event that "*hath already been*"– in our universe.

In this figure, the now non active diagonal event lines cut across the (equally non active) vertical evolutionary (history) lines. An event line can represent all the events of the world happening simultaneously or just a Now local snapshot in the simplest case. Here, I take the event line as representing a local snapshot because K specifically used the key word *Now*. So in verse 15 it is appropriately taken to refer to a snapshot. Note that in the figure I have shown (as usual) only a skeleton array of vertical lines representing universe histories. In reality each line should be viewed as being representative of a huge bundle of universes.

This demonstration of steps can also be made to work using only two universes (taking only U_6 and U_8) and still appear to address verse 15. That would also apply for people who lived in the universe adjacent to the Template, but in chapter 5, I argue that we must be far enough away from the Template to allow precognitive dreams containing recognizable motion to occur and be recognized. That would therefore require a large number of universes. Next, I will present a decoding of the second challenge.

"*The Four Quartets*"

If you were not so impressed with the way E 3(15) was dealt with, you should be more impressed with an interpretation of a T. S. Eliot (TSE) masterpiece of poetry that helped earn him a Nobel Prize in literature. Here is the unexpected connection to Eliot: while scanning through back issues of the *Journal of Scientific Exploration*, I discovered an invited essay[15] that was a discourse on time written by Professor Terzian of Cornell University. At the very end, the author in obvious desperation threw out this amazing quote from TSE, whose reflections on the weird nature of time, he thought, was best summed up by this stanza.

Or say, that the end precedes the beginning,
And the end and the beginning were always there,
before the beginning, and after the end.
And all is always now.[16]

By quoting this, Terzian implied that his subject matter was in a state of extreme confusion (which was true) and that he was unable to unscramble it at least in the short space available to him, so he seemed to be saying in closing that even a Nobel Prize in literature could also get confused with time. This was my introduction to Eliot's philosophical poetry involving time.

In answer to a question about the interpretation of his poetry, he had declared that it was not amenable to translation. I decided to find out if either of the two propositions (1) that Eliot was apparently confused about time and (2) his poetry was beyond translation or was true or falsifiable.

Terzian may not have realized that Eliot was highly intuitive and maybe had an undisclosed geometrical sketch he had made. Did he use a modified version of Dunne's model of time published only seven years earlier? This is possible but highly unlikely as it would have required a considerable amount of further work to make it even partly comprehensible. Besides, Eliot would hardly have missed the opportunity to publish his version of a corrected geometrical scheme explaining time.

In chapter 4, I give a short biographical sketch of Eliot that shows his proximity to Dunne, but I can see the influence of A. N. Whitehead in TSE's spatial explanation of time. As in the first challenge, I will again be using a section of figure 2.1A. Because of the particular way TSE phrases his verse(s), I use the beginning and the end of World War II as the two events. Figure 3.2 shows two snapshots of the Blueprint; the first one shows us at the beginning of World War II and the second one just comfortably after that war.

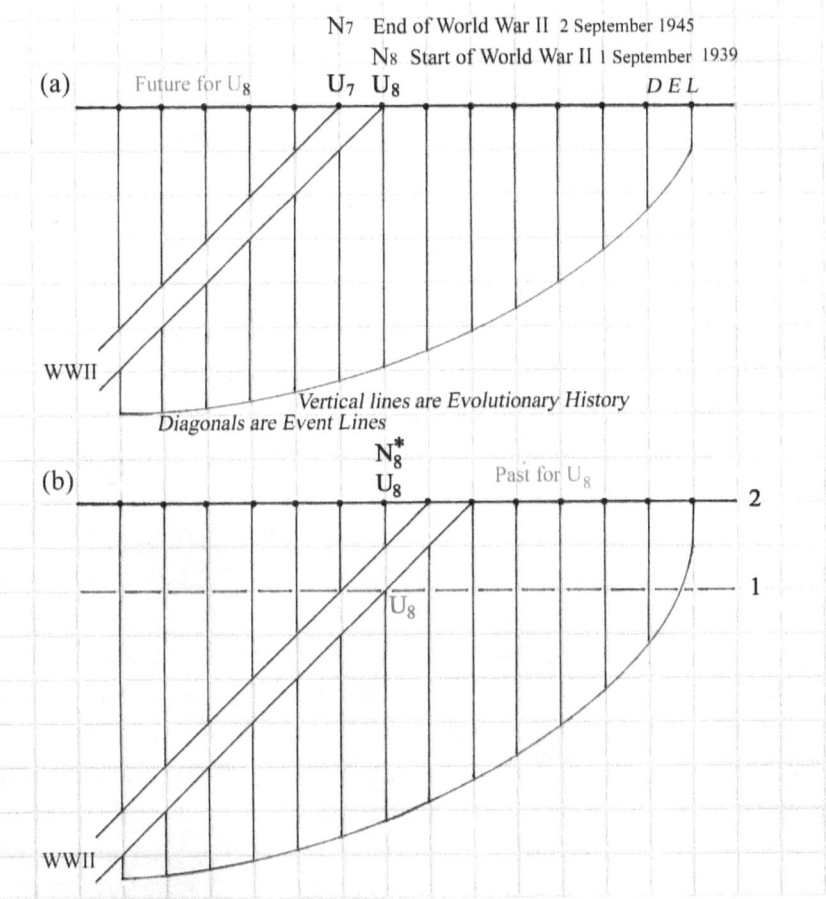

N7 End of World War II 2 September 1945

N8 Start of World War II 1 September 1939

(a)

Future for U$_8$ U$_7$ U$_8$ D E L

WWII

Vertical lines are Evolutionary History

Diagonals are Event Lines

N$_8^*$

(b)   U$_8$ Past for U$_8$ 2

U$_8$ 1

WWII

Fig 3.2 is based on a cut-out of the Blueprint. **(a)** Inside U$_8$ the snapshot day is September 1, 1939; Britain declares war on Germany. World War II has begun (first event). From *outside* the multiverse, we see that *the universes ahead of us (packed between* U$_7$ & U$_8$) contain the *complete saga* of WW2 *ending* at (second event) N$_7$ in U$_7$, where the date is September 2, 1945. From a perspective *outside the multiverse* these two dates happened simultaneously. U$_7$ is *ahead* of U$_8$ and moving left; the COPI procedure 'moves' events to the *right*. In Figure **2.1B** the universes move clockwise & the COPI procedure moves counter clockwise. Thus, the *"end"* of the war (in U$_7$) *"precedes the beginning"* (in U$_8$) satisfying **Line 1** of TSE. In **(b)** the *DEL* has shifted up from position **1**-as in **(a)**- to **2**. The extended Event Lines intersect the new *DEL* and preserve the relativity of the two events. So, as in **Line 2**: *"the end and the beginning were always there"*. It would be more appropriate to use *'are'* in place of *"were"*. **Line 3** (a corollary of line 2) is confirmed, and **Line 4**: *"all is always now"* poetically expresses *'eternalism'*. In **(b)** I use N$_8^*$ in DEL position **2** to indicate, that the 'Now' there is different from the 'Now' in position **1**. It could also be expressed as N$_{8,j}$ where j is some huge number.

Some who can recite "Burnt Norton" may ask, "What about the first five lines of "Burnt Norton"—how does that translate?" I ask them to wait until chapter 5, in which I tackle that task. Interested readers are encouraged to attempt a translation themselves but should be prepared for a surprise.

I surmise that TSE might have struggled to understand the second part of Dunne's book in which he attempted to build his model of time but ended up creating an infinity of discrete times occurring within one universe, which is clearly impossible. At that stage, Dunne should have realized he was in need of a serial multiverse. If you don't feel you immediately want to try to dissect Dunne's time diagram (figure 8 in the third edition of his book), do not worry because I do a substantial part of that in chapter 4.

Is it also possible that TSE was incorporating concepts of the special theory of relativity into his verse? I think not as he was clearly focusing on events in Newtonian space. How then did TSE do his figuring? That is the big question. In the absence of an Eliot diagram, which he might have sketched out but put in the waste paper basket to make it look as though he had written this passage out of his head, TSE seems to have grasped the essence of the relativity of events. That is essentially what is contained in chapter 2.

Next, I present the last of the three challenges. For this one, it is imperative to modify descriptions containing the word *time* by replacing them with expressions denoting motion, change, events, or evolution.

Zimmerman's Challenge

American philosopher Dean Zimmerman exposed a fundamental philosophical problem I have taken up as a challenge to further probe the strengths and weaknesses of the serial multiverse theory. This problem is centered on two ways of looking at time as presented by the English philosopher J. E. McTaggart[17] in 1908. In an essay[18] concerning a philosophical analysis of McTaggart's A and B series of time, Zimmerman asserted, "No one, to my knowledge, has defended the following combination of views: time lacks an intrinsic direction but [yet] includes objective distinctions between past, present, and future." This proposition contains two conditions I denote as 1 and 2. The proposition appears to border on a paradox. That is the challenge to be resolved using the STZM.

Many readers may already be able to respond to this challenge. I will present my response to it using the concepts covered in this book. Recall that time, denoted by t, is a scalar quantity, a numerical value obtained from a clock, so we are in type-2 time. It can be a single recorded Now moment value read off in hours, minutes, seconds, or fractions of seconds. If we use a stopwatch, we obtain a scalar time interval denoted $\Delta t'$, with a prime to distinguish it from the special and fixed value Δt shown in figure 2.1A.

What follows is for the benefit of non-technically trained people. The main use of a clock is to monitor an event. Thus, if we are monitoring movement of objects, the motion will be referenced to three mutually perpendicular axes (x, y, z). Let movement of an object take place parallel to the x-axis. This is a special case just as Einstein specified for starting his theory of special relativity. Measurement of a distance is required; call this value Δx. This distance was traveled by the object in duration $\Delta t'$. Now, if you are not interested in the direction of that line, Δx, the speed of the object is simply $v = |\Delta x / \Delta t'|$, a scalar quantity. But if you are interested in direction, the line drawn becomes a vector and an arrow head is put on the end of it. Thus, the velocity, v, given by $\mathbf{v} = |\Delta \mathbf{x} / \Delta t'|$, is a vector quantity. However, $\Delta t'$ remains a scalar quantity. Therefore, time lacks an intrinsic direction. The arrow relates to the direction moved; t' goes along for the ride. Thus, condition 1 is upheld.

The proposition above requires that condition 2 includes objective distinctions between past, present, and future without conflicting with condition 1. When viewed inside our universe, time is traditionally tensed by referring to it as past, present, and future (P_1, P_2, F), but we should be using the following terms: $P_1E \equiv$ past events, $P_2E \equiv$ present events, and $FE \equiv$ future events, which are still relative terms. Events are intimately connected with rates of motion that contain scalar time used in addressing condition 1. But the word *objective* in condition 2 would hardly be met. *Objective* would need to be replaced by *relative*, and hence, condition 2 would become subjective. Thus, Zimmerman's proposition would be seen as not having been met. This is where the STZM model is able to provide some support.

Our only recourse seems to be to view the universes from an external viewpoint, which is easy for our consciousness to achieve. Refer to figure 2.1B. Focus on the universe labeled You Are Here, which is our universe U_8 (containing N_8) marked in figure 2.1A. It is in the current Now moment. If then a conscious entity is able to have an out-of-universe experience (OUE), which is equivalent to just viewing the diagram right now, the observer will see what appears to be an objective basis for assigning P_1E, P_2E, and FE *labels* to the universes.

Assign P_2E to occur in U_8. Then, by the convention of the diagram, all the universes to the right of N_8 (U_8) would be given P_1E labels and those to the left would have FE labels. In figure 2.1B the tense assignment is FE clockwise of the location 'You are here' (P_2E) and P_1E below it. The principle of cause and effect is not violated. Thus, the terms of Zimmerman's challenge appear to have been met.

This exercise introduced me to McTaggart and his belief in the unreality of time. It can now be stated that both his A series and B series of time are valid views as long as we couch the argument in terms of events and specify our view location. The serial multiverse scheme is the only way I know of responding to this challenge. Next is a famous and famously flimsy paradox.

The Grandfather Paradox[19]

I end the chapter with this classic paradox that can be shown to be absurd by use of the multiverse model. There have been other dismissals of the paradox, but this is the only one that uses a geometrical multiverse model to do so. Use of the word *time* here is rather hard to get rid of, but I use it only sparingly when I know that readers should understand what I mean (that is by understanding chapter 2). In the STZM model, time is defined in each universe as a universal clock time, and thus it is available for tracking objects in a space. Clocks can be used anywhere in each universe. Rather than speak of time travel, I say that fundamentally one should use the term *space travel*. The time machines of H. G. Wells and M. Crichton are suitable only for a museum of science fiction; figure 3.3 and its caption shows how I deal with the paradox.

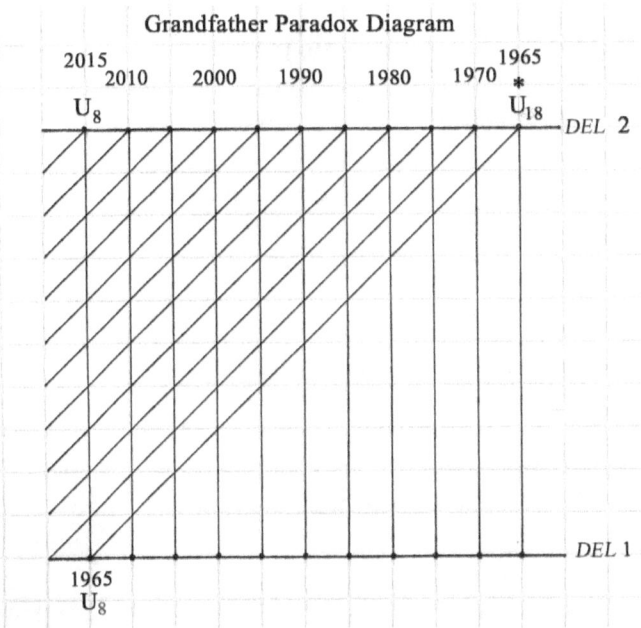

Figure 3.3 (A cut-out from the Blueprint). Our protagonist is living in 2015 within universe U_8. He and the family have a huge grudge against his grandfather, who died five years before. His ashes lie in the local cemetery. Slide *down* the *diagonal* line, from 2010 label on the horizontal *Dynamic Events Line (DEL 2)*, until directly below U_8. That is in the 'dead' past domain–as are all the other vertical and diagonal lines. Our protagonist is a frustrated physics philosopher, intrigued by the idea of time travel. *If it was possible to travel back to 1965*, he surmises, *before my grandfather was married–and kill him–that would mean that neither my father nor I could have been born. Now there is a paradox for you! But given that I had a fast spaceship, fuel and sustenance as well as the navigation equipment, it looks as if, in principle, it could be done.* So he takes a spaceship and arrives in the year 1965 on planet earth in U_{18}* (*not the one* in Figure 2.1A). A search for his grandfather is successful. Our protagonist raises his rifle but in an instant a spirit says in his mind: "*Don't shoot! That isn't really your own grandfather! He is a clone of him*". (The diagonal event line running *up* from U_8 to U_{18} is where the COPI procedure operates). "*Besides, he has another soul which is different from your grandfather's soul.*" "*Now you*

ought to get back home". "As before, the return trip is through about 1.57 x 10^{15} –or more than one and a half million, billion–universes, *but the trip will take longer".* Why is this? Shouldn't the Grandfather Paradox be just struck off the record? But wait…what *if* the clone *was* shot? It is of interest to think this scenario through. That will finally convince you of the farce that it is and show that you understood the COPI procedure.

Notes

1   That was the experience of John W. Dunne, who is briefly mentioned in chapter 1 and featured in chapter 4. He was a pioneer in the attempt at understanding the nature of time. The main value of his book *An Experiment with Time* is the documentation of a selection of Dunne's dreams along with some others and his analyses of them. They demonstrate that the human consciousness has an ability to roam in space and in time zones that are displaced from our clock times. Many of his dreams qualified as being precognitive, but others couldn't be *a priori* placed in that category. Dunne made an unsuccessful attempt to model time geometrically in this same book and in others that followed.

2   L. Dossey, *The Power of Premonitions* (USA: Dutton/Penguin, 2009).

3   R. E. Guiley, *Harper's Encyclopaedia of Mystical and Paranormal Experience* (New York: HarperCollins, 1991), 467: "[18] of the [39] books [in the Old Testament] are ascribed to prophets" and contain prophecies. The famous dream of Pharaoh is considered a prophesy according to the interpretation of Jacob and the ensuing events. It was a static image; these are usually symbolic. This story illustrates the fact that some people are unable to interpret their own dreams. Others have run this Bible count; see note 4.

4   D. E. Dean, *"Precognition and Retro-cognition,"* in E. Mitchell and J. White, eds., *Psychic exploration: A challenge for science* (New York: Putnam, 1976).

5   The SPR and ASPR would make it appear that precognition is just as difficult to prove as the other preternatural phenomena, but I am not a supporter of this view.

6   Wikipedia has a short article on this phenomenon. Considering that the term was introduced by the French psychic researcher Émile Boirac in 1917, it is significant that the article is being frequently edited or updated as monitored from January 18, 2012, to November 25, 2016. Other Internet sites are useful to consult especially on the use of the comprehensive term *déjà vécu.*

7   D. H. Powell, *The ESP Enigma: The scientific case for psychic phenomena* (New York: Walker Books, 2009). Powell, an MD, referred to work reported and new terms introduced in a book by P. M. H. Atwater entitled *Future Memory.*

8 http://en.wikipedia.org/wiki/Reincarnation. The lead paragraph contains a misleading description of reincarnation. It is neither generally a religious concept nor a matter for philosophy. It is a spiritual concept and it should be judged to transcend and become independent of both of those areas. For another reference, see www.jewishencyclopedia.com/articles/14479-transmigration-of-souls. This is an unedited version of the *Jewish Encyclopaedia* published between 1901 and 1906. Others have already been through the exercise of identifying how many references to reincarnation there are in the Bible.

9 Dr. Ian Stevenson, vastly underestimated, was foremost in this research amid continual skepticism. His collaborator was Dr. J. Tucker, who continues research in obtaining memories from young children. Over a decade beginning in 1975, Stevenson wrote four volumes on "Cases of the Reincarnation Type" published by the University of Virginia Press. They were based on his interviews of young children living in India, Sri Lanka, Lebanon, Turkey, Thailand, and Burma. Other countries included the United States (Alaska), and countries in Africa. These are really cases that fit reincarnation when no other reasonable hypotheses can be identified.

10 B. L. Weiss, *Same Soul, Many Bodies* (New York: Free Press, 2004). Weiss is an MD in psychiatry and uses directed life regression techniques.

11 A case illustrating this is http://reincarnationstudies.com/anne-frank/.

12 R.A. Monroe, *Ultimate Journey* (New York: Doubleday, 1994); P. Elder, *Eyes of an angel* (Charlottesville, VA: Hampton Roads, 2005).

13 D. H. Stern, *Complete Jewish Bible* (Maryland: Messianic Jewish Publishers, 1998).

14 E. S. Christianson, ed., *Ecclesiastes through the centuries* (New York: Wiley, 2007), 70. Quote taken from http://en.wikipedia.org/wiki/Ecclesiastes. According to J. B. Priestley, "Both Thomas Wolfe and William Faulkner are clearly time-haunted men." There were many more, some of whom are included in chapter 4.

15 Y. Terzian, "The nature of time," in *Journal of Scientific Exploration* 2(2), (1988), 143–54.

16 Taken from part V of "Burnt Norton," the first of the "Four Quartets" written ca. 1934 by T. S. Eliot.

17 J. E. McTaggart, "The Unreality of Time," in *Mind,* 17 (1908): 457–74.

18 D. Zimmerman, professor of philosophy at Rutgers University, http://fas-philosophy.rutgers.edu/zimmerman/Presentism%20and%20Rel.for.Web.2.pdf

19 The so-called grandfather paradox appears under different guises but the basic structure is essentially the same.

CHAPTER 4

Overview of a Special Era of Time Philosophers

Overview

In this chapter, I occasionally have to use the generic, loose, unqualified word *time* in the discussion. John Dunne occasionally used it that way, the philosopher-mathematician Alfred Whitehead also admitted this practice for the same reasons. Because Dunne was the first person to try to nail down the problem of time in a way that no one else had evidently done before, I extensively discuss his book *An Experiment with Time* first published in 1927. In it, he postulated the existence of a bewildering, seemingly infinite number of serial times somehow related in a fictitious way. In this chapter I identify how Dunne created such a series of times that lacked locations, except the first two.

A few years after the book appeared, perhaps between 1929 and 1933, Whitehead, without any mention of Dunne, identified two types of time: a formal time (or common astronomically determined clock time) and a nonformal time, which was not well defined initially but was evidently based on remarkable events embedded in a formal timekeeping system. This is traceable to Greek ideas about time that seem in principle related to the North American Indian winter counts, which are based on recording outstanding events throughout a year. They were later to serve in calibration of the time series.

William Hammerschmidt pointed out that Whitehead later shifted his definition of nonformal time to mean something quite different. It took a later philosopher, Richard Feist, to reinterpret Whitehead's terminology and suggest that his nonformal time should better be called metaphysical time.

According to the Blueprint in chapter 2, a system of timing is essential to the operation of the multiverse. I call it type-1 time, but it could also be described as metaphysical time. This suggests it may be equivalent to Whitehead's nonformal time, which is far from explicit. Rather than being fully transparent as formal time is, this newly revealed timing system in the STZM shows only a facet of its probable nature. It is based on a time pip at a regular interval denoted in chapter 2 as Δt.

Serious practical scientists and almost everyone else consider formal clock time as the only time. In his book *About Time*, Paul Davies quoted Sir Hermann Bondi as writing that time was a human construct. This is consistent with Whitehead's formal time. The identification of a nonformal time requires an excursion into the realms of a multiverse. In chapter 5, I conclude that there is a manifestation of this second time in our human experience. The concept is discussed in terms of a quantum computer clock. This may be recognized by focusing on the possible nature of the Now moments as Einstein was encouraged to do later in life. However, it was an obscure Frenchman, J-M Guyau, who in the late-nineteenth century already offered a clue. Much later, Whitehead offered another insight but still couched in philosophical thinking.

Yet given only the existence of formal time to think about, most people still become confused by its real nature. To get around this, I have shown in later chapters that from a purely semantic standpoint, only word selection is required to fade clock time into a background quantity (a mere number) by referring to descriptors such as events, change, and motion, thus temporarily freeing us from the influence of the perceived power of clock time. In *The End of Time*, J. Barbour had in principle done this.

Dunne features rather prominently in this chapter even though his model of time (based on precognition) is definitely incorrect but with a curious spin to it. He and I unconditionally accepted the observational evidence that the ability to see future events is neither an illusion nor a hallucination despite there being many strange aspects to it. Two of the most puzzling aspects are the unpredictable nature of its occurrence and why it occurred in the first place. After reading Dunne's book, I followed his protocol for identifying a precognitive dream, and I succeeded after three years. I had entered what I considered to be a significant dream in my dream logbook on the morning of the April 10, 2013, but I soon forgot about it.

From June 19 to July 12, prolonged and disastrous flooding hit Calgary, where I used to live. Extensive stretches of the Bow River overflowed from Canmore to Calgary. I still did not remember my dream of seventy days earlier. It was only when I was later checking my logbook for details of another dream (number 10), that I found dream number 8, which matched the rising waters of the Bow River. But there was subtle symbolism contained in the dream suggesting that the video-like images had been rendered somehow resulting in deviations between the dream and reality. The effect is like a smoke screen. Where had that been done?

In *Power of Premonitions* (2009), Larry Dossey, MD, provided us with another example. In his first year of medical practice, he experienced a week of premonitions about patients all of which came true. He had never had them before; they seemed to have come out of left field. After the occurrences stopped, "it was as if the universe, having delivered a message, hung up the phone." Though appearing quite distinct, it seems that precognitions and premonitions are closely related in that they share the same pathways. Dossey appears later in the next chapter.

Significantly, several twentieth-century researchers were involved in acquiring numerous accounts of precognitions by experiment, interviews, or correspondence[1]. These included Camille Flammarion, R. L. Mégroz, J. W. Dunne, John B. Priestley, and more recently, the *New York Times* columnist Robert Nelson. This practice has not been maintained, justifiably I believe because it just proves the same thing multiple times. The claim that less than 1 percent of a large population experiences true precognition is I believe an underestimate because of the excessively restrictive guidelines for authenticity imposed by such authorities as the Society for Psychical Research (London) and the American Society of Psychical Research (New York).

Strange as this may sound, I think the phenomenon was not intended to occur yet it pervades our existence, which seems to be dominantly directed toward living and concentrating on the physical life. It seems that recipients of precognitive dreams invariably do not know why they get these dreams and do not know how they appeared to them. Dunne was a good example of such a recipient. He was however motivated to try to find out. This indicates to me that precognitions and premonitions are very likely flaws in a system designed to make our existence look identical to the one in the Template.

This flaw view seems to be essentially consistent with what theoretical physicist J. D. Barrow concluded; he said as much by writing that our world contains flaws and reasoned, "The flaws of Nature are as important as the laws of Nature for our understanding of true reality."[2] This has to be seriously factored into our thinking when making an overall assessment of this book.

Typically many of the late nineteenth-century and early twentieth-century scientists-turned-philosophers largely worked alone. This was certainly the case for Dunne, whose study of time was in a sense heavily myopic. His range was restricted to his own perception of precognitive dreams and those of his friends and relatives whom he later experimented with before starting to write a book about his particular and rather unique study of time.

Those working in this subfield of preternatural phenomena work under the cloud that their data are non-reproducible at the case-by-case level. It is for just this reason that mainstream, classical scientists avoid this area of inquiry; they have to contend with the weird and almost magical aspects of general relativity and quantum mechanics.

The purpose of this chapter then is to look mainly at the twentieth century and note the activities that took place in humanity's quest to understand time. For example, by hacking into Dunne's weird time model, I opened up some new ground. I have found it very revealing and instructive to first place Dunne in the context of his contemporaries because it is evident he was to some degree in the beginning quite influenced by some of them but later widely diverged from the then-mainstream views.

The principal characters contemporary with Dunne are time-haunted men, a term due to J. B. Priestley, who was on his own list of such persons. I considered myself on the list for a while but no longer. I have tried to order my short list roughly in chronological sequence though overlap and informal interaction between them makes it difficult to determine the origin of some of the key ideas that emerged in the late nineteenth and twentieth centuries. You (the psychologist) will be able to judge if these are normal or abnormal people.

The Main Characters

Charles Howard Hinton (1853–1907)

British mathematician Charles Hinton pursued eclectic interests, which suggests that by today's standards, he was a kind of polymath. While teaching at a college and a school in England, he studied at Oxford University and obtained a BA in 1877 and an MA in 1886. He married the daughter of George Boole, who established the principles of Boolean algebra, a branch of mathematical logic especially applicable in computer coding.

He left England with his wife and spent time in Japan before moving permanently to the United States. In 1893, he was instructor in mathematics at Princeton University. He soon moved to the University of Minnesota, where he was an assistant professor until 1900. While there, he designed and built an elaborate automatic baseball pitching machine that required a good grasp of the geometry of trajectories and dynamics. His last two jobs were at the U.S. Naval Observatory in Washington, DC, and the U.S. Patent Office, where he acted as examiner of chemical patents.

Even before 1880, Hinton had explored the geometrical depiction of four spatial dimensions seemingly as a way to associate time with spatial dimensions. To help visualize and keep track of the dimensions, he devised a set of colored cubes. Such thinking revealed his geometrical insightfulness as well as his inventiveness. All his publications, which had appeared by 1904, tackled the problem of space and time as being inseparable, thus in a sense paralleling Einstein's work but more explicitly Hermann Minkowski's geometrically and mathematically expressed concept of space-time. Such local system thinking is not conducive to discovering the multiverse.

A decade before, H. G. Wells's book *The Time Machine* had appeared. In 1905, Einstein's article on special relativity was published. This concept was in principle not new but rather a stunning clarification of existing ideas. Time was in the searchlight of physics. Clock time as rendered by special relativity was suddenly seen to possess some special but puzzling properties that affected the shape of physical objects in space. These developments were perhaps propitious for many other people thinking about time anomalies.

One of these individuals was Dunne. After experiencing a considerable number of dreams with subsequently verified precognitive content (such as from newspaper articles or by personal experience), he took on the daunting task of figuring out the proper way to look at time. He was attracted to Hinton's ideas on the subject and studied them in connection with designing a purely geometrical model of time. In addition, he either interacted with or read the publications of a number of people who appear in this chapter. This did not lead him to the multiverse either.

Jean-Marie Guyau (1854–1888)

The inclusion of Guyau[3] here may come as a surprise to some readers as it was to me because he is largely unknown today. Guyau was a precocious student, and at a very early age—about twenty—he was appointed to teach philosophy at the celebrated Lycée Condorcet in Paris.

One of Guyau's pupils there was Henri Bergson, who later appeared due to the nature of his writings to be following and expanding on topics Guyau had earlier written about.

In 1879, Guyau resigned his academic position to seek a more favorable climate in the Mediterranean because of a rapid decline in health. Even in his last years, he had still managed to write influential works on various philosophical topics. He died young due to a pulmonary disease, likely tuberculosis, but not before completing an important manuscript, *La Genèse de l'Idée de Temps*, published posthumously. Some commentators claimed it was based on a purely mental representation of time, but this is hardly so if J. A. Gunn[4] said Guyau was opposed to the psychology of time. That he recognized time as separate from space is explicit: Gunn wrote, "Time is measured by reference to space, but we nevertheless regard it as distinct." This is broadly Newtonian physics thinking but he should have written that time is used to measure the rate of movement of objects in space; that is, it monitors events.

However, his belief in the transient event nature of our existence is explicit in this statement: "The world is seen to be a world of events, coming into being and passing away." This indicates that he was aware there was something special about the sequence of events. There is a suggestion here that broadly thinking, he had similar thoughts as Einstein had later in life about the Now moment and that something he needed to understand was just beyond his ability to grasp. It appears to me that Guyau was brushing close to the concept of cosmic time just as Friedman and Einstein were in the early twentieth century. This will be made clear in the next chapter.

Guyau certainly introduced something very significant when he wrote in *La Genèse de l'Idée de Temps:* "We misconceive the future if we think of it as that which comes towards us; it is rather that towards which we go." This is based on the inside the universe view as shown in figure 2.2a. In another view as seen in chapter 5, I show graphically that it is possible to imagine that as our universe moves forward in the torus, in a sense, we do approach the future. But simultaneously, our new Now physically comes in from the universe ahead. It is clear that neither Guyau nor Whitehead showed any signs of viewing our universe from outside it. It is a surprise that Whitehead was able to make his seemingly counter statement on the matter: that the future comes toward us. I deal extensively with this matter in the next chapter.

Elsewhere, Guyau seems to have written with a different spin as in, "Time is a simple product of consciousness … Time … is … a kind of systematic tendency, an organization of mental representations."[5] Thus, contrary to Gunn, I do recognize there is a psychological component to Guyau's thinking as well as a metaphysical one although here it is only the latter in which we are interested.

Henri-Louis Bergson (1859–1941)

Bergson, of Polish origin, lived briefly with his parents in England. The family later moved to France, where he attended Lycée Fontanes in Paris from 1868 to 1878. Bergson won a prize for his scientific work and another prize in 1877 for the solution of a mathematical problem published the following year in *Annales de Mathématiques*. Rather than following a career in the sciences as expected, he was somehow influenced in favor of the humanities. At nineteen, he entered the famous École Normale Supérieure, where he obtained the degree of Licence-ès-Lettres and Agrégation de philosophie in 1881.

His dissertation (translated as *Time and Freewill*) was submitted along with a Latin thesis on Aristotle ("Quid Aristoteles de loco sĕnserit") for his doctoral degree, which was awarded by the University of Paris in 1889. *Time and Freewill* was published the same year. He was opposed to the belief in predestination. He thus believed that the future was not real and therefore could not exist.

In 1901, he wrote a very important essay, *Introduction to Metaphysics*, which was useful for others in understanding his major works that covered seemingly diverse fields such as physics, biology, and evolution. He became a good friend of William James, the American philosopher (referred to as the father of modern psychology), but they had a fundamental disagreement in the free will and determinism debate. However, it appears that Bergson had influenced James enough at that time to shake the latter's faith in the power of pure logic. This as expected was where Bertrand Russell (a strict logician) was quick to find fault with him. I agree with Russell.

At the turn of the twentieth century, the theoretical physics community included two already prominent figures: the Dutchman H. A. Lorentz and Frenchman Henri Poincaré. In 1905, the stage was set for a major paradigm shift in physics that further complicated the understanding of time in local domains in which the speeds of objects are extremely high. The development of Albert Einstein's theory of special relativity was heavily influenced not only by the work of Lorentz but also by Einstein's own job in the Bern patent office reviewing patents dealing with synchronizing clocks. But special relativity has strongly counterintuitive aspects that shook the logic (or commonsense) aspect once again.

Bergson's critical stand against the power of logic seems to be consistent with the fact that in May 1913, he was comfortable enough to accept the presidency of the (British) Society for Psychical Research (SPR), and he delivered to the society an impressive address: "Phantoms of life and psychic research." It could be surmised that he was open to this area of inquiry, which seemed to defy logic. His main presence in the SPR, however, may have been in keeping with the largely monitoring and data collecting capacity of that organization and to detect any fraudulent activities in research into the supernatural.

In 1914, the Roman Catholic Church put three of Bergson's books on its index of prohibited books. Unlike his much younger English counterpart, Bertrand Russell, Bergson was not a pacifist when it came to World War I. Ten years later, he considered converting from Judaism to Catholicism, which seems ironic enough bearing in mind the Vatican's decision against him in 1914. However, there were other compelling circumstances in Europe at that time to be taken into account. In 1922, Bergson ineffectually attacked Einstein with the publication of *Duration and Simultaneity: Bergson and the Einsteinian Universe*, which was followed by a polemic with Albert at the French Society of Philosophy. It was largely for the book *The Creative Evolution* that he was awarded the Nobel Prize for literature in 1927. Bergson is often hard to gauge as he ranged over a large spectrum of fields.

Dunne mentioned Bergson's ideas early in the development of his model of time. According to Dunne,[6] Bergson (at that time) seems to have remained steadfast in his belief that the fourth dimension is "a duration of events" rather than being encapsulated in the explicit term *time* in the sense of Hinton's or Minkowski's model, where it was assigned for mathematical reasons a dimension of length. The notion of duration of events is complicit with something in space being measured by a stopwatch to get a scalar quantity in seconds. I have already addressed this topic in chapter 2 and will expand upon it again in chapter 5, a repository for discussing many enduring and related topics. It is interesting to note that the concept of duration in time had already been advanced in the fourth century BC.[7]

Bergson was explicit in stating that there could not be a preexisting future and thus that we all automatically had free will. Some today still believe this, but from the model presented in chapter 2, it is clear that free will can occur only in the Template, and we are definitely not in that universe. Dunne enigmatically claimed in his first book that his own model "accounts [for] self-consciousness and freewill." So on that point, Bergson and Dunne seemed to have been on the same page, but this is inconsistent with Dunne's belief in prophesy, which can only indicate the existence of strong determinism in our lives. He evidently attempted to squeeze some evidence from his own dream accounts for the existence of an element of local, self-generated free will.

I suspect that he would have approved of the term the *Dunne effect*, which is a new and tangible term involving soft free will, introduced in chapter 5. It is needed to cure a paradox.

So Bergson was correct in his thinking if he was living in the Template universe, but he had no knowledge of any universe other than the one he was in. Thus, he lived under the illusion of free will. Bergson's association with the SPR would have put him in contact with an archival database of precognitive dream information (dating from 1882). This record would have appeared by then to have had a moderating effect on disbelievers. However, there were skeptics in the SPR (or at the very least those who demanded the most rigorous proof and protocol in investigating any paranormal phenomenon), and of course, the range of paranormal phenomena studied then had very noticeably broadened.

Alfred North Whitehead (1861–1947)

Whitehead, who was awarded the British Order of Merit and who became a fellow of the Royal Society in London, was an influential British mathematician, logician, and philosopher in that order. As a philosopher, Whitehead seemed ahead of his time; he had considerable influence on American philosophers of a later generation especially physicist-philosopher David Bohm.

Born in England and educated at Cambridge University, he quickly reached prominence in the fields of mathematics and logic and later process (and depth) philosophy. Though he was a contemporary of Bergson, I have not found any animosity between the two such as occurred abrasively between Bergson (the belligerent) and Einstein. With the exception of his well-known and close collaborations with Bertrand Russell (producing the three-volume *Principia Mathematica* in 1910, 1912, and 1913), Whitehead seemed to have worked mainly in his own introspective and self-motivated world. He produced a version of Einstein's relativity theory and made it available to Einstein, who seems to have put it aside. I originally planned to avoid Whitehead because of his excruciating way of presenting ideas until I discovered he considered mathematics to be a subset of logic, but more importantly that he identified two types of time, and he subsequently had some most essential and enlightening interpreters.

After 1911, Whitehead turned primarily to issues of the philosophy of science and ontology. In *Process and Reality* (1929), he maintained that the flux of things (i.e., the events involving movement) is basic to our philosophical thinking. This was also Guyau's view. He further said that nature had to be thought of as "a structure of evolving processes" that formed the reality. This is evidently the source of the term *process philosophy* with which he is identified. In this book, you will see that working with events rather than processes is the most profitable course to take for reasons that will become obvious.

Whitehead came to my attention after I had written chapter 2, in which logic was the basis of my setting up the geometrical scheme, and in which I targeted events as being the main elements. He appears later in chapter 5, in which I deal with his time metaphysics quite extensively with the help of two very lucid translators.

The next character is Dunne, who gets a disproportionate amount of coverage for reasons that will become apparent. He seems to have missed out on the wisdom of Whitehead though there is oblique evidence that the reverse situation doesn't seem to be entirely missing. But Dunne's detailed documentation of precognitive dream images was evidently lost on Whitehead.

John William Dunne (1875–1949)

Overview

Dunne was originally the main reason for writing a chapter of this type. I will expend more words on Dunne because he set out on a determined solo mission that lasted three decades to hack the concept of time using a geometrical scheme. There are three main locations devoted to Dunne in this book. This one is in the form of a short but very enigmatic biography. The second one is in chapter 5, and a concluding location is in the form of an essay in chapter 6.

To obtain sufficient biographical material, I had to hunt through several of his books published after *An Experiment with Time* in 1927 to obtain sufficient information about him. The book yielding the most personal information was *Intrusions?*[8] which was his last one. It was made possible by his wife, Cecily, Marion Violet Joan Twistleton-Wykeham-Fiennes (take a breath) Dunne. She was the eldest daughter of the wealthy eighteenth Baron Saye and Sele. Dunne's early life is revealing and seems to have set him up for writing his enigmatic but popular first book. This was because of the first half which dealt with an analysis of selected dreams.

Biography

Dunne was born in Ireland, the third son of General Sir John Hart Dunne and Julia E. Dunne, Anglo-Irish aristocrats. When John was six, he suffered a serious undisclosed accident that essentially confined him to a "mobile bed for three years"[9] and limited him to crutches for two more years. During this time, he exhibited a "precocious frame of mind" (bordering on the philosophical) and even incredibly conceived the outline of serialism in the nature of time. This suggests that he might have had precognitions of his future at that early age, and this turned out to be the case.

His mother considered him "too smart for his own good." She often rebuked him for his "stubbornness." It is also very revealing and relevant to learn that between ages twelve and thirteen, he "experienced on three or four occasions the curious thing,[10] which is called ... ecstasy." Then, "Inspired by a Jules Verne story at the age of 13" (about 1888), Dunne "dreamt of a flying machine that needed no steering, that would automatically level itself regardless of wind or weather."[11] Such a machine, the D.8, was realized later near the end of his aeronautical career. So this apparently was Dunne's first recorded precognitive dream significantly (as will be seen later) not recorded in his 1927 book.

The full extent of this man's education is not at all clear. He attended the exclusive Eton College, the choice of royalty and wealthy families that produced several prime ministers. This comes from his 1927 book in which he experienced a terrifying event involving a wasp that had crawled inside his "Eton collar." He presumably entered Eton at age thirteen. In his last book, he wrote that at age seventeen (1892) supposedly clear of high school, he "was a pupil on a South African farm"[12] east of Cape Town. This was just after the first Boer War in which his father had served. The senior Dunne was later appointed an emissary by the British government and based in Cape Town.

During that time, the younger Dunne did a self-styled psychoanalysis and concluded that he was "in some strange way two diametrically different persons occupying the same body" and that he was aware that he "could easily choose which one of the person types he wanted to be." To satisfy the "aggressive personality," he "learned boxing."

A year or two later, he decided to investigate spiritualism. This involved experiments with automatic writing and attendances at the Cape Town Spiritualistic Society, which held public séances. About that time, he joined the Cape Town Chess Club, which he thought might moderate his developing curiosity in the area of psychic phenomena.

At that stage, when he was eighteen or nineteen, there is no mention of his entering a university. He later reminisced on his career writing that he "grew up to be a soldier and a pioneer of aviation."[13] He thus downplayed his parallel career in metaphysics, which most of his books clearly project. I suppose he took that to be obvious and just wanted to stress that he did at one time have two respectable jobs. In 1927 he suppressed his involvement with earlier spiritual involvement for fear that it might affect sales of *An experiment with time*. He began studying aeronautics while on medical leave from the army in England in 1902. Encouraged by H. G. Wells among others at that time, he designed and built a number of small test aircraft models based on a swept-back biplane wing configuration.

By 1904, after another short commission in the army, he was ready to proceed to constructing gliders and eventually powered aircraft to embody his concepts of flight control and stability based on the gliding properties of an exotic African seed pod.[14] From mid-1906 to 1909, he was assigned to the new Army Balloon Factory in South Farnborough. While there, under strict security, Dunne built an experimental manned glider, the D.1, with provision for fitting engines and propellers later. These activities of course were running parallel or slightly lagged aeronautical developments primarily in the United States, Europe, and other centers in England with which Dunne was not initially connected.

Dunne, the Rising Aeronautical Engineer
I can only briefly mention his aeronautical career.[15] Much of his designing was evidently empirical being based mainly on experimentation marked by rather expensive prangs (a British and colonial term for nonlethal crashes). Two of the main features of his flying-wing biplane incorporated the swept-back leading and trailing edges and the positive to negative incidence of the top wing's leading edge toward the tip.[16] Out of his ten models, the D.8, was a success; it corresponded closely to his dream at age thirteen.

As World War I was looming, the aeronautical community in England had decided that aircraft should be designed around a robust fuselage separating wings and a tail assembly rather than based on a 2-tier wing as Dunne had done. He had built a basic fuselage-design airplane, but it was not a success; neither was D.10, his last creation. He became a fellow of good standing in the Royal Aeronautical Society, London. You can thus see that Dunne was very flexible in his activities, but he was an experimentalist, not a theoretician.

Dunne, the Dreamer and Architect of Serial Time
As the originator of serial time, however, the hypothesis that followed proved to be fatal. Like Hinton, Dunne displayed the qualities of a self-motivated and largely self-trained polymath where his math was

represented by geometry just as in the STZM model. His later math involving equations of relativity is quite irrelevant and would have been a signal to physicists that he was clearly making an incorrect link by connecting unrelated dots.

After his first book was published in 1927, Dunne received attention from the Society for Psychical Research (SPR) in London. He was told his dream reporting did not conform to a sufficiently rigorous protocol. One of the requirements was for a witness to be present when he wrote down his dream description and have it immediately read and signed by the witness. The witness should hold an exact copy. His wife, Cecily, could have done this, but I feel sure the SPR wanted an impartial witness.

In about 1930, T. Besterman coopted Dunne and some university students in an experiment to test the fidelity of precognitive dream ability. Besterman acted as the readily available witness (which he ought not to have done as you can learn later by reading about Jung and Pauli's early situation). The results were published in an article in the SPR journal. Out of seventeen of Dunne's dreams occurring over four months, Besterman stated that four were "suggestive of precognition."[17] This should be considered a very good rating considering that Besterman had the backing of the SPR and that "a score of 23.5% was 25 times the national average."

Henri Bergson was president of the SPR twenty years earlier, and as was then known, he was quite critical of metaphysics. He had written a book on the subject. He could have established the critical tone of the SPR and for good reason. It took me one and a half years to capture just one dream that was, I might say, very strongly suggestive of precognition even with symbolism involved. I was acting without a witness, so the SPR would not have accepted it under their excessively strict protocol. I did not feel it was justified waking a neighbor at 3:00 a.m. to get a signature on my dream report. It was only because the dream was a rather unusual one that it was logged because it did not obviously contain an urgent message.

The empirical approach to Dunne's flying-wing designs of one and a half decades earlier seemed also to apply when he was building his scheme representing time. He was it appears his own designer of the time diagram, writer, and reviewer. Whereas his years of designing airplanes set him up for the geometrical element of the time diagram, I judge that the earlier years resulted in a sort of myopic thinking that robbed him of the chance to make a breakthrough into the multiverse domain against which he was already strongly biased. It would have served him better if he had given his manuscript to qualified people before sending it to the publisher.

He acknowledged his faithful wife, Cecily, for having typed the final manuscript and a supportive unnamed acquaintance at the University of London for "continuous encouragement." Quite understandably, there is no overall support from Sir Arthur Eddington, with whom he briefly corresponded. There was apparent agreement from Eddington, however, that a form of serial time must exist, but Eddington did not elaborate on this.

Dunne in Decline

For Dunne's last book, an editor (possibly T. S. Eliot, then at Faber and Faber in London) is mentioned but no reviewer except for an acknowledgment by John's widow, Cecily, to a Reverend Dr. Ahrendt, for "correcting the proofs" of the mathematical appendix. It contains elements of relativity theory introduced into England at the time by Arthur Eddington. This theory was not at all relevant to Dunne's time model.

Dunne had evidently termed unannounced visits by what he called a familiar angel as an *intrusion*. It was a nonmaterial holographic vision of the type commonly seen by dying people. John's son (John Geoffrey Christopher) wrote a brief account of the fourth intrusion experienced by his dying father, who was then seventy-four. This time the angel was accompanied by a "raging tempest." It was "pitch black," but he could sense what he discerned to be the angel's "white robe" (having seen him before under less-disturbing circumstances). The elder Dunne had said "that he thought rapidly for some question to ask the angel." Then, the question that had always worried him came out: "Christianity—is it true?" The immediate reply was, "God lets it be true for those who want it to be true." The son said that his father "had no interpretation of the reply."

My first reaction to this anecdote was one of great surprise. I would have expected Dunne to have asked the angel this question (again in the common English way of speaking at the time, i.e., subject first): "My theory of time—is it true?" In this case, would the above response from the angel have been expected? My tongue in cheek interpretation of what the angel said is that God does not interfere with humanity's affairs in copied universes simply because he knows that in this universe, our lives are by design predetermined. Sic fiat.

The Dunne Effect: A Tribute to the Man and Solving a Paradox
I discovered at an advanced stage of writing this book that a paradox existed in the front end of the multiverse model when I applied a necessary test to it using Dunne as my test subject. I realized that in the Template, the first John Dunne could not have dreamed of the future because there was no future available. Because Dunne's life there was devoid of any precognitive dreams, he therefore had no reason to write *An Experiment with Time* in the form we know it. He would still have been equipped to dream about the past, but if he did, it might not have seemed to be very remarkable and not worth a book. In any case, no such book exists. It cannot be assumed that he would be alone in dreaming into other universes just as is the case in this universe.

So the first copied universe covers all his non precognitive activities only. As the number of universes increased, the copied Dunne (same body, different soul for each universe) would start getting short glimpses of the future until in one universe, say the billionth, the billionth Dunne would be dreaming of complete events that would subsequently be actualized in his universe. As a result of this, his life took on a new turn. He was suddenly motivated to work out a theory of serial time (which we know he had conceived of even as a boy) and to write the book: *An Experiment with Time*. It was the first one to appear in the multiverse.

A special simplifying case would be if one assumes that this happened in our universe. If it happened in a future universe, it would eventually be copied to us. As I discuss in the next chapter, I hypothesize that the process might involve human energy levels that can override the deterministic component of our existence. These could correspond to what we call *eureka* moments. This will be especially appealing to those who want to claim every bit of free will they can.

So you can see that this effect is one our Dunne could never have known about. This you will discover by carefully rereading *An Experiment with Time*. It was only by thinking of him and his activities in a mind experiment (held in the STZM) that I was able to present the explanation needed to mitigate the paradox.

In the above I have drifted into the sometimes controversial modern beliefs that are bestowing great and unexpected potential powers on humanity. This stems mainly from an interpretation of quantum mechanics theory and experiment and specifically on the very profound result highlighted in chapter 5 that our physical existence involves connectedness[18] throughout the universe. This was a basic belief of David Bohm and a fundamental property of the serial time-zoned multiverse. In every COPI step, all the contents of a universe have to be copied together. We are all included with everything else on this gigantic file.

Recall that Dunne[19] wrote something that impinges on the above: that we are included with everything else in this universe but that humans are not afforded priority over the rest. This is where we fundamentally differ. Dunne argues for the clinical approach; I would argue for the anthropocentric approach because as I conclude in chapter 5 the multiverse was evidently designed around the process of transmigration (incarnation) of spirits into physical bodies.

Herbert George Wells (1866–1946)

Wells is here first because of all the other people in this lineup, he was probably most closely connected with Dunne and was intensely interested in the time problem himself even before he met Dunne at Farnborough. Second, he is well known as a writer, even a pioneer, of science fiction and speculator of future political and technological events. He authored thirty nonfiction works including *A Text Book of Biology* (1893). He had studied biology taking courses from T. H. Huxley, who led him to his interest in organic evolution. He also took courses in physics though his bachelor's degree was in zoology. This education would have brought him in close contact with the one important subject of his era: *time*, particularly in its connection with the rates of progression of biological changes and because of its great, emerging importance in physics.

Because he was also writing on the evolution of technology, Wells was also particularly interested in aircraft development in the early part of the twentieth century and made the acquaintance of some of the best-known aeronautical engineers including Dunne, who was a profitable source of information for some of Wells's fictional technology books. He even used Dunne as a character in one of his books, and he later used Dunne's stories of dreaming the future as an element in another futuristic novel, *The Shape of Things to Come* (1933). Reading about Wells and his book titles suggests that he might have experienced, as had Dunne, dreams of the future. However, it is not clear how much benefit Dunne derived from his association with Wells over and above using him as a sounding board to try out his ideas on the subject of the new serial time. From 1914 to 1927 (excepting the interval 1915–17, when he was a munitions instructor for the army) Dunne was presumably working on the manuscript for his first book.[20]

During those thirteen years, Wells wrote fifteen novels, twenty-four nonfiction books, five short stories, and four articles all of which were published by 1927. Of course, the depth of each man's subject was entirely different. Two of Wells's books directly concerning time were *The Time Machine* (1895) and *The Conquest of Time* (1942).

In addition, Wells was trying to hatch a scheme using some of Dunne's aircraft patents to attract funding from military contractors at the time. After writing a complimentary review of *An Experiment with Time*, Wells much later told J. B. Priestly that he had become disenchanted with Dunne's serial-time hypothesis.

Carl Gustav Jung (1875–1961) and Wolfgang Ernst Pauli (1900–1958)

These two very special characters were brought together in 1932 because of Pauli's need of counseling on marital problems he was having. I have put them together here for reasons that will become obvious. Jung, a Swiss, was a prominent member of the analytical depth psychology élite; Pauli, an Austrian, was a prominent representative of the European theoretical physicist élite. They formed a long and profitable collaboration in dream analysis and on archetypes without a major rift developing.[21] My interest in Pauli is twofold: firstly he was a prolific dreamer who in 1934 had a spectacular dream that has a striking similarity to fundamental geometrical aspects of the Blueprint diagram in chapter 2 and secondly his realization that it was necessary to consider consciousness and preternatural phenomena in conjunction with established science to arrive at an integrated understanding of our existence.

The Jung-Pauli interactions endured from 1932 until a few years before Pauli's death in 1958 and resulted in two joint books: *The Interpretation of Nature and the Psyche* and *Atom and Archetype*. Jung personally experienced some very strange, preternatural phenomena as well as some precognitive experiences. Coincidently, he was born the same year as Dunne, so they had been exposed to the same general atmosphere and attitudes toward unusual and unexplained phenomena. There is reference to Dunne's 1927 book in Jung's collected papers, but there is no indication Pauli ever read or commented on any of Dunne's books. If this is so, I find it not at all surprising.

Jung practiced analytical psychology and psychiatry in Zurich and used material gathered from his patients to formulate explanations for some of his well-known psychological theories and paranormal experiences. A vast amount of material (including anecdotal dreams and their interpretation) formed the basis of several of his books. Despite his being viewed as a mystic, he actually strove to be formally scientific in his publications. This, mind you, after he was reported to have said he was "rather disappointed over [his] father's academic approach to religion." My interest in Jung is heightened because (as of this writing) I have experienced two long, serial sequences linked by a common symbol[22] as well as a closely occurring minor one in 2010.

Pauli's interactions with Jung began when the former was suffering deep psychological problems. His father suggested he get counseling from Jung, who also lived in Zürich. The problems had developed from the collapse of his first marriage in 1929, the later death of his mother by suicide, and his father's immediate remarriage to a woman Wolfgang's age. As a result of these family crises and poor communicative and social relations with his academic colleagues, Pauli had sunk into a habit of spending his evenings at bars, which led to his reputation of exhibiting inappropriate social behavior. At this time, Pauli was a professor of physics at the Swiss Federal University (ETH) in Zürich, and his deteriorating social behavior was becoming unacceptable to his colleagues at ETH. Jung had a clinic in the city and was to be given a professorship on the faculty at ETH the following year.

Jung wrote that during his first meeting with Pauli, "he felt the wind blowing over from the lunatic asylum."[23] On the one hand, he recognized immediately that at the heart of Pauli's problem was his difficulty in relating to women. On the other hand, he also importantly recognized that Pauli would make an excellent test subject (with the object of being a source of data for his next publication on dreams). He referred Pauli to a younger colleague who was an analytical psychologist so he could be an impartial witness and evaluator only of the data.

It turned out that Pauli provided hundreds of dream accounts, which was very fortunate for Jung. He had already written a paper on dreams in 1909 (the year he had first met Freud) and was contemplating a major assault on the subject [24] when Pauli came to see him. He had permission to use the dream material but had strict instructions not to reveal its source.

It seems that by 1934, Pauli had been subjected to enough therapy that he was able to adjust to another marriage and continue a successful career in physics with better control of himself. Following this, the two men maintained a close relationship through regular visits and then through letters. Pauli allegedly provided Jung with about 1,000 dream descriptions though apparently Jung considered only 400 of those of interest. I do not know how many of these dreams were precognitive and how many archetypal or a mixture of both types. Jung also incorporated quite a number of his own dreams into the analyses. Thus, the relationship between Pauli and Jung developed into one of considerable mutual interests because apparently they were both to differing degrees and orientation motivated toward understanding the nonphysical side of life.

Pauli deliberately concealed his relationship with Jung from his colleagues in the physics community for fear of being ridiculed. Thus, he was a sort of Dr. Pauli and Mr. Hyde. He was notorious for his severe criticism of theories or theses that were badly presented or untestable and thus unfalsifiable. He viewed the worst of these as being not able to be evaluated and not properly belonging in the realm of science even though they posed as such. They were, he said (in the most forceful translation) worse than wrong because they could not even be proven wrong.

On the recommendation of Einstein, who mentored him, Pauli was awarded the 1945 Nobel Prize in physics. His contribution was called the Pauli *exclusion principle*, which applied to electrons in orbits around a nucleus in atomic theory. This aspect of Pauli, normally operating in a mainstream physics mode, contrasts widely with a bizarre preternatural phenomenon he was also involved in.

Occurring locally, this repeating phenomenon was a very puzzling near coincidence between Pauli's presence and some nearby destructive event. It was named by a colleague the *Pauli effect*. It typically occurred when Pauli entered a university laboratory in which there was sensitive monitoring equipment in operation. They would spontaneously malfunction. It applied even to other objects in other surroundings such as a vase of flowers falling over as he entered a reception room in a building in Zürich.

Once, a colleague in Germany wrote to Pauli saying the equipment in his laboratory had inexplicably ceased operating at a certain time on a particular day and asking where Pauli was. Pauli wrote back saying that he was on a train that had stopped at the colleague's railway station at that time. In all probability, Pauli was thinking of that colleague at the same clock time. Is this an example of the power of human thought? Is there another agent involved?

This is hardly an isolated phenomenon. Jung could produce a very similar effect; he described one in his last book, *Memories, Dreams and Reflections*. It happened famously in the library of Sigmund Freud. There were other people known to be able to spontaneously produce the same type of effect and still others who could produce the effect on demand as in a planned demonstration.[25]

As mentioned earlier, Pauli had one dream that is of considerable interest in connection with the Blueprint in this book. It was sketched from Pauli's dream records and described in an article by Roth.[26] In the dream, "a man resembling Einstein has drawn the following figure." The diagram is shown redrawn here as the graphic in figure 4.1a. Pauli wrote, "It showed me quantum mechanics and so-called official physics in general as a one-dimensional section of a two-dimensional ... world, the second dimension of which could only be the unconscious and the archetypes." That was all. He later changed his mind about the identity of the man; it may not have been Einstein after all, but that does not affect what follows.

It is possible that Pauli was significantly influenced in his interpretation of the dream diagram by his acquired knowledge of Jungian archetypes; it is also possible that I am influenced by my familiarity with the Blueprint, which the essence of the Pauli diagram strongly resembles. I hopefully suggest that the whole diagram is a previously undescribed archetype composed of elements of the Cosmic Blueprint diagram and a general symbolism of the embracing fabric of quantum mechanics. I specifically involved the many interacting worlds (MIW) model because it shares properties with the STZM.

(a)

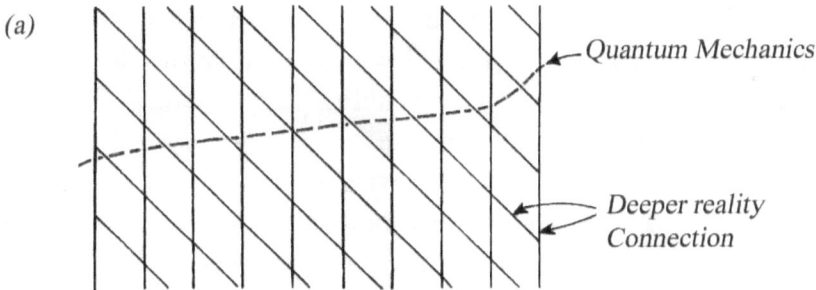

Fig 4.1(a) Redrafted version of the diagram that Wolfgang Pauli dreamt in 1934. The figure of a person on the right *with his back to the viewer* is unnecessary. Pauli seemed to interpret this as a symbolic modular diagram representing an intimate association between **quantum mechanics** (the rough slanted upper line), "official physics" (**vertical** lines) and the "The deeper reality connection" ("domain of the archetype and the unconscious within life") as the **diagonals**. The lower two labels are entirely my interpretation (see text).

What I am particularly struck by in this diagram are the two sets of straight lines that share relative orientations with the lines of evolution (verticals) and the event lines (diagonals) as seen in the Cosmic Blueprint except that the array is horizontally flipped 180°. Actually, an early draft of the Blueprint did have the diagonals sloping the same way as in figure 4.1a due to my taking the cause-and-effect progression along the *DEL* in the other direction (left to right) equivalent to moving the start of the GEs (the primary event line) to the right side of the Blueprint.

Although Dunne's 1927 time diagram (his figure 8) has a partial similarity with Pauli's dream diagram, I maintain that figure 2.1A of this book takes priority. This is because Dunne's diagram lacks a set of diagonals not because they couldn't be drawn in but because he seemingly didn't realize this was significant. Moreover, his vertical lines should not have been extended below the single diagonal one. That was an error. This action of Dunne's moving horizontal line also occurs in the *DEL* of the Blueprint, but in that diagram, all the actualizations are occurring in the dot universes that make up the horizontal line as it moves upward.

The transverse dashed line, labeled Quantum Mechanics in figure 4.1a appears to have been added last as if to indicate it is embracing the other two sets of lines suggesting a symbolic connection between quantum mechanics and, quoting Pauli, a "deeper reality" (viz. a multiverse). The MIW multiverse and the Cosmological model in this book provide this deeper reality. Pauli learned about subconscious-existing archetypes from Jung. In their 1955 joint publication *The interpretation of nature and the psyche,* Jung wrote about his theory of synchronicity while Pauli chose to write about Kepler in the context of an archetype.

The upshot of this exercise involving Pauli's dream diagram is that the whole of figure 4.1a may be regarded the result of a double contribution to an archetypal symbol to which Pauli was privy. If he was the first Pauli to dream of that diagram, it would be regarded as a special case. This is sufficient to get the idea across. There is no difficulty if an earlier Pauli (in a universe ahead) had the first dream of the archetype, but it had to have happened well behind the Template, according to the Dunne Effect. It would be passively copied along to our universe via the COPI procedure. The symbolic dream obviously qualifies as a precognitive one.

In the special case, it has only just recently been actualized following the last two proposed multiverse schemes STZM (2008) and MIW (2014). Jung's apparent lack of commenting about Pauli's interpretation of figure 4.1a implies that he was unfamiliar with it. If this is an archetype, it would evidently be associated with a synthesizer and a philosopher. Just who created it and where is a mystery just as is the mystery of who created Jung's famous eight-element fish symbol sequence and who created the six-element feather symbol in my serial, significant, linked-events experience. This last term should also be applied to Jung's fish symbol sequence and others like it.

Figure 4.1b is a simplified rendition of a part of figure 2.1A (the Cosmic Blueprint). I have added to it the upper, dashed, crosshatched lines above the horizontal line (the *DEL*) to make it look more like Pauli's diagram except that it is still a mirror image of it. The upper dashed lines indicate that they have not yet been actualized in any of the universes along the *DEL*. Each solid diagonal line can be envisioned as being composed of a stack of identical old Now moments (like an ultrafast snapshot of part of an event). It remains unclear why the diagonal event lines might have turned out to be sloping left in Pauli's diagram. In this respect, both Dunne's diagram and the Blueprint are consistent.

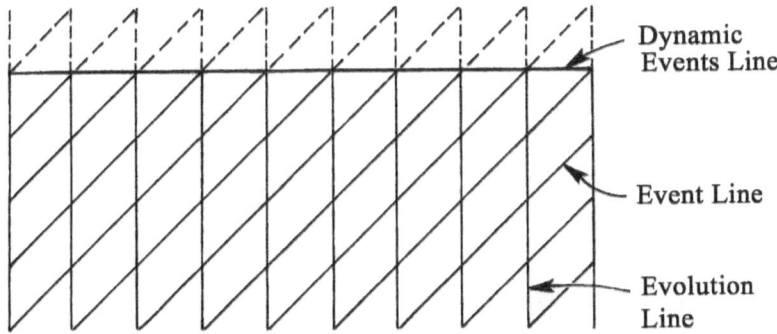

Fig 4.1(b) This diagram represents a cut-out rendition of Figure 2.1A (Template) with the addition of upper dashed *event* and *evolution lines* that have yet to be 'actualized' in *all the universes* represented by the current Dynamics Events Line (*DEL*) shown by the bold horizontal line. That line acts as a scaffold for the multiverse, which must contain quantum mechanics. It represents the concept of *eternalism* (eternality) when animation is applied, stasis is broken and the line moves upward. The *DEL* is not in Pauli's diagram. Instead there is a cord-like line which he explicitly labelled 'quantum mechanics' as shown in Figure 4.1(a). The lines below the DEL are all past events (represented by the diagonals) and historical trajectories (represented by the vertical lines).

Figure 4.2 is a result of modifying, simplifying, and demystifying John Dunne's figure 8 in *An Experiment with Time* (3rd edition, 1958), so the few essential features that correspond to the multiverse Blueprint are made obvious. The diagram serves another purpose that will be discussed after identifying the main features of it.

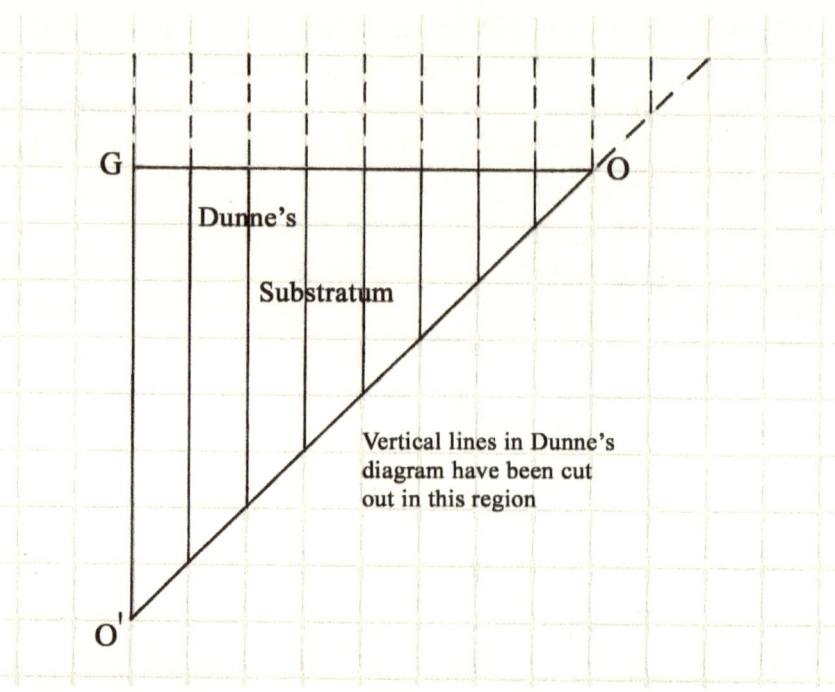

Fig 4.2 I will first point out where certain parts of the original Dunne diagram in *An experiment with time* have been cut: (1) a section of vertical lines in the area indicated were deleted because they are incompatible with the STZM and with Dunne's other specifications (2) the extension of line G–O to Dunne's point H (not shown here) was removed. Like the Blueprint, Dunne's diagram is dynamic. Line O′O is fixed and corresponds to the primary event line of Genesis Events in the Blueprint. Line G–O moves continually upward and corresponds to the Dynamic Events Line (*DEL*) in Figure 4.1(b). The vertical lines extending below GO to meet O′O (the substratum area referred to by Dunne) correspond to the vertical Evolution Lines in the Blueprint. Dunne's *"entities"* are located along G–O; they are the dots (universes) in figure 2.1A. The features relevant to the multiverse model that are missing are the complete set of diagonal lines that would (as in figure 2.1A) populate the rest of the 'substratum' area. Upper vertical dashed lines as in the previous figure are equivalent and refer to future actualized events in each universe. Diagram is shown in stasis so line G–O corresponds to the frozen *present*; all the vertical lines of the substratum represent the 'dead' past for a person living in any one of the "entities" (universes) that populate the line G–O. The diagram contains *two* times, *as explicitly labelled by Dunne*. One belongs in the vertical axis; the other belongs in the horizontal axis (not shown). By correspondence with figure 2.1A, Dunne's vertical axis 'time' would be the 'timing system'– not the second member of Dunne's 'serial time', which is fallacious. The proper 'time series' of clock time in each universe exists along G–O. The reference to John Dunne is his 1927 book *'An experiment with time'*. In subsequent editions he changes the numbering on his diagrams.

I have proposed that in this special case, our Pauli (or his subconscious, or the collective unconscious) supplied a preview of figure 2.1A and later combined elements of that image with a one-dimensional symbolism of the MIW model when it appeared to be related to the STZM model as shown in this book. It took eight decades for this two-component archetype to actualize in this universe. This can all be modeled using the architecture of the STZM.

I now discuss how Dunne may have been involved. Again, this can also be modeled using the STZM architecture. Could he have dreamed of figure 2.1A? He was definitely known to have recalled several notable precognitive dreams as related in his first book. Other persons of note are reported to have had dream assists in their research; even the composer Franz Schubert had dream assists which he concealed in a document that was not to be opened until 50 years after his death. Let us suppose that Dunne dreamt (possibly somewhere between 1918 and the early 1920s) only a fleeting glimpse of figure 2.1A fully assembled by 2008. What did he probably remember of it? What did he make of it? Here is a list of possible elements of his thinking process.

1. He captured the triangular form consistent with the orientation of figure 2.1A.

2. He recalled the set of vertical lines that became his substratum.

3. He recalled some diagonal lines but was unable to figure out their function except that he saw how to utilize one diagonal line that appears in his figure 8 (in the third edition of *An Experiment with Time*); it forms the hypotenuse of the triangle.

4. In the above figure, he referred to "O'—O" as a diagonal but didn't know how to prove that it sloped at exactly 45° though he drew it that way. As seen in a later book by him (*The Serial Universe*), he saw a diagram that expressed special relativity in a geometrical construction due to Hermann Minkowski.

There, the light cone sloped at exactly 45°. He evidently used this value, but in so doing, he misappropriated it. Special relativity is completely irrelevant to the present problem.

5.   He knew that the line G-O moves up; this is equivalent to the dynamic events line moving up in the Blueprint (as seen in the animation of it). Such an animation was completed in July 2013 just before the MIW model was published.

6.   Dunne seemed to have acquired just enough geometrical components to be able to put this list of elements together to enable him to form his figure 8 and to make it look as if it would work according to his singular mind-set about serial time. The leap in complexity from his figure 7 to his figure 8 with only an obtuse descriptive text between them is a signal that he already had the basic diagram for figure 8 in his mind. He seemed to be missing the crucial intermediate steps shown in figures 2.1a, 2.1b, and 2.1c of this book. Then, because of his determination to cram everything into one universe (instead of a multiverse), he was left with a completely useless result. It was a classic case of so near yet so far.

Because this second-guessing of Dunne's thinking is speculative (particularly to those who have not studied his books), it is hardly worthwhile to pursue the matter any further than this. But it shows one way how some scientists may acquire information and speaks to the assertion that some of us in the present may be quick connected to some of those in the future. Pauli showed this ability.

Because the Pauli dream diagram is well documented and reasonably well constrained, I will demonstrate using the Blueprint the pathways that carried the image to Pauli and how it was copied into our universe shown as of 2014. The diagram in figure 4.3 displays the geometry of this. It shows the two positions of the dynamic events line (*DEL* 1 and 2) as of 1934 and as of 2014 in our universe, U_8. The information pickup in 2008 when figure 2.1A was available is not labeled, but it is held in memory and combined with the MIW quantum mechanics component in 2014 represented by the sloping dashed line in figure 4.1a.

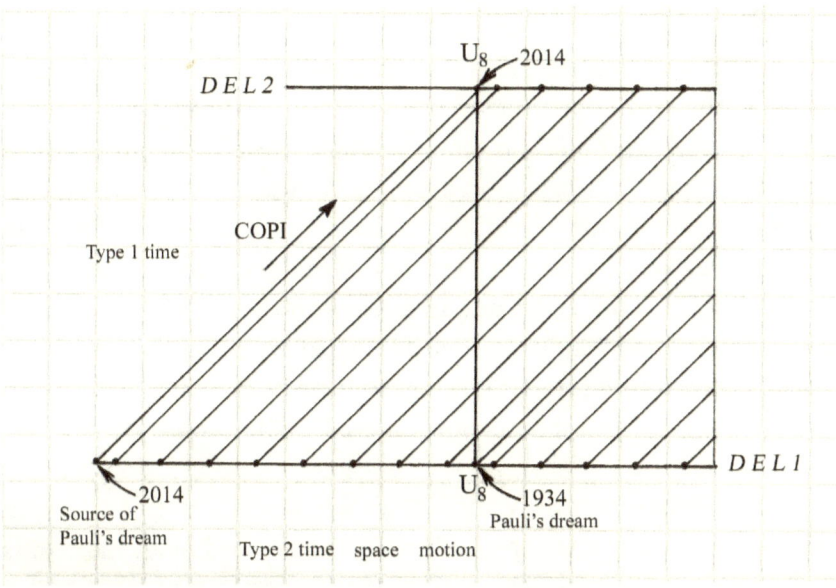

Fig 4.3 This diagram is based on the Blueprint. Directional reference frame axes are not shown as time has no direction. In the up-page direction Type 1 time, as in the Blueprint, denotes the Δt spaced time PIP's that control the motion of the Dynamic Events Line (*DEL*) as it moves incrementally upward from position *DEL1* to *DEL2*. On the lower horizontal span Type 2 time is contained *within* universes (black dots); Type 1 time is also there *between* universes with an *imprint inside* them via the Now moment. Copying is continually taking place along the diagonal event line as shown by the COPI arrow. Pauli's dream occurred in 1934 within U_8. It was only a snapshot Now moment image perhaps existing already in the collective unconsciousness as an Archetype. It was captured in Pauli's dream memory and reproduced in hard copy for Carl Jung. This was when the STZM and MIW models were in conjunction in the Blueprint at a much earlier 2014 time (see lower left corner of the triangle). I now hypothesize that the Archetype was synthesized by some unknown process in a spirit environment, but also involving Pauli's subconscious. Because the decision by me to build the STZM was based on anecdotal accounts of precognitive dreams, the 'Dunne Effect' must be invoked. The COPI routine incrementally transported the dream image (in hard copy) up to the dot marked '2014' in the *DEL2* position in U_8 after a duration of 80 years. Only the one (vertical) 'evolution line' is shown highlighted to reduce confusion. Recall that in actuality there are dots between dots...etc.

Figure 4.3 is the special case in which Pauli of our universe is taken to be the first Pauli to dream the snapshot seen in figure 4.1a. It is an idealized composite of two widely separated events (the dream and the actualization). It should be represented by a simple animation to avoid confusion, but I will attempt to step through this in what follows. Refer to figure 2.1A to see a complete Blueprint diagram. Visualize the lower dynamic events line (*DEL* 1) (position 1) when it had nothing above it and was rendered static. Recall that our universe is U_8. Pauli has his dream sometime in 1934 (setting our resolution in this diagram to one year). The universe to the far left labeled 2014 contains the event that led to a symbolic amalgamation of the STZM geometry (of this book) created in 2008 with the MIW model based on a quantum mechanics model published in 2014. This symbolic image is shown in figure 4.1a. It had to have been synthesized by an intelligent entity, and it somehow entered Pauli's consciousness via his subconscious.

The *DEL* is now slowly moved up perpendicular to its length. Simultaneously, the COPI procedure leaves a record of all the events of each universe all the way up the 45° diagonal event line until it is stopped for inspection when the year 2014 occurs in U_8. Here as earlier, you can see how our present comes from the future just as Whitehead posited in 1933. All the way along that event line, each universe contained a physical body looking just like our Pauli but with a different soul.

I have made reference to people such as Pauli, Whitehead, and others many times not because they were necessarily the first to broach an idea but because their various legacies are still very much with us. Our next character had an entirely different approach to explaining time. His ideas were expressed in a coded metrical form. One of his stanzas of the "Four Quartets" was analyzed in chapter 3. The second stanza, of less accuracy than the first but with some instructive value, is dealt with next.

T. S. Eliot (1888–1965)

Thomas Stearns Eliot is included in this chapter; even though he was not one of a group of technical philosophers seeking to understand time, he demonstrated he had grasped the basic concepts conveyed by the word *time*. It was in an advanced form for that era even though clothed in poetry. On first reading, TSE's time appears to be aligned with McTaggart's B series of time, which is compatible with Cosmic time explained in chapter 2. That immediately puts it into a multiverse context. This was already alluded to by the second challenge in chapter 3. However, TSE chose not to formally throw his hat into the technical philosophers' ring but instead chose to express his understanding of time in a different and enduring genre of human communication: poetry.

To his credit, however, TSE had the right background for the task. He had taken courses in anthropology and philosophy at Harvard University and had earned a BA in three years. He must have had a solid grounding in logic in the philosophy courses. He studied philosophy at the Sorbonne and attended lectures by Henri Bergson. In short, Eliot gave the impression of being a polymath, which is a good position to be in when it comes to tackling a slippery subject of this type.

Though he was an American by birth, Eliot decided to live in England, where he attended Oxford University. He later taught Latin and French at English schools. He was also knowledgeable in ancient Greek and German. In 1916, he completed a doctoral dissertation in philosophy at Harvard but decided not to defend it; that would have required traveling by ship to the United States during World War I. Only the year before, a German U-boat had sunk the Cunard passenger liner *Lusitania*. In 1925, he joined Faber, a well-known London publishing company that in 1927 became Faber and Faber Limited. Eliot had many of his later books published there. As we have seen, that company published Dunne's first book, *An Experiment with Time*, in 1927. It is a reasonable assumption that Eliot might have had a hand in editing that book. I surmise that he would have viewed it mainly from the aspect of a philosopher. As a matter of fact, Dunne, an engineer, had actually been referred to as a philosopher, but it was really analytical metaphysics he was dealing with. However, as I understand it, metaphysics seems to overlap naturally with philosophy and has tended to become part of it.

At about this time, some private correspondence took place between R. L. Mégroz and Eliot. Mégroz (of London) wanted to know whether Eliot had any interesting dream anecdotes to provide him for inclusion in an anthology of dreams he was preparing. Eliot replied that he did not experience interesting-enough dreams and that Mégroz should obtain the recent book by J. W. Dunne. It turned out that Mégroz had already done that.[27]

The following is relevant to material in chapter 3. "Burnt Norton," the first of the "Four Quartets," is the source of the quote analyzed there. It is contained in section V, lines 10–13 of that quartet, which was written between spring 1934 and fall 1935 and published in 1936. Dunne's second book, *The Serial Universe*, was published in 1934. To find a documented link between Eliot and Dunne, I searched the books that dealt with the life and times of TSE. It was quite revealing to read in the book by Canary,[28]

> Although recognizing the Bradleyan colouring of Eliot's critical writings, Wollheim[29] observes that Eliot fed on less and less substantial fare, so that second-rate theology and middle-brow books (like J. W. Dunne's An Experiment with Time) may have much more to do with the metaphysics of the Four Quartets than Bradley or any other philosopher ... Recognizing that Eliot eventually repudiated Bergson, Smidt[30] was one of the first to call attention to the influence on ... TSE, of *An Experiment with Time*.

However, my reading of "Burnt Norton" is that TSE paid more attention to the philosopher Alfred North Whitehead, from whom he must have grasped the profound realization that in this universe, the present moment comes from the future enunciated about 1933.

Besides, already by 1929, most philosophers and physicists had largely discredited Dunne's model of time. The fact that TSE claimed he had never had any interesting dreams seems to rule out the possibility that he had a preview of the Blueprint geometrical model, which would have provided all the material he needed. But then, McTaggart's 1908 article might also have given him some strong hints as I have already identified.

In the preface of Bergsten's book,[31] that author quoted TSE as saying, "Poetry is a constant reminder of all the things that can only be said in one language, and are untranslatable." But, all languages (including poetry) have some rules (however loose), so there must be some way of translation; otherwise, poetry is just a fancy way of arranging words and of no practical use, just nice to listen to when read like some minimalist music.

Eliot took his time writing "Burnt Norton" as he was involved in what turned out to be a highly successful drama, "Murder in the Cathedral" at the time. He was asked to remove some material from the script, so he relocated some of it to "Burnt Norton." The latter thus seems to have had an irregular evolution spanning several years. This may have had something to do with the disparity between the discourses on time at the start of "Burnt Norton" (section I, lines 1–5) and the stanza that occurs in and was analyzed in chapter 3. So bearing in mind I believe that *time* defined here is relatively confused compared with the stanza analyzed earlier, let us review how "Burnt Norton" starts.

> Time present and time past
> Are both perhaps present in time future;
> And time future contained in time past.
> If all time is eternally present
> All time is unredeemable.

The second, third, and fifth lines of this stanza do not make sense. To be compatible with the right thinking behind the stanza in section

V of "Burnt Norton," the complete stanza here should be fixed so that it reads in accordance with the STZM model. I will temporarily retain (for the sake of poetry) the bad habit of using the raw word *time*. In so doing, I am able to eliminate a line. There is a limitation though—the stanza applies only to copied universes.

> *Time present and time past*
> Are both derived from time future.
> *If all time is eternally present*
> All time is redeemable.

It is rewritten below more acceptably and still in TSE's style by avoiding use of the word *time* altogether without actually killing the clock.

> *Events present and events past*
> Are both derived from events future
> *If all events are eternally present*
> All events are accessible.

I replaced *unredeemable* with *accessible*. We know that because the Blueprint suggests it and our experiences of the many seen past and future events (whether in dreams or visions) verifies it. The disparities in this first stanza and the higher fidelity of the second variation on a theme of time do not support TSE's claim that his poetry is untranslatable. Moreover, errors (typically inconsistencies) can be detected and corrected. It is left as an exercise for the reader to modify the above stanza so that it is valid in the Template, but it may need another line.

Notes

1   N. C. Flammarion, *L'inconnu* (*The Unknown*) (London: Harper & Brothers, 1900). His collection of accounts was acquired before the turn of the twentieth century and was on the order of 1,000 anecdotes. The French were then

very active in paranormal research. The London-based Society of Psychical Research was established in 1882 and started archiving paranormal accounts including precognitions. R. L. Mégroz was active in the mid-1920s and was researching for an intended *Anthology of Dreams* as reported in the *London Times Supplement*. He is referred to briefly under the subsection on T. S. Eliot. After years of hunting, I have located his book *The Dream World: A Survey of the History and Mystery of Dreams* published in 1939 by Random House in London as well as Dutton in Boston. There is a recent second edition published by Kessinger in Montana.

In 1963, John B. Priestley, author of *Man and Time* (see appendix 4) acquired over 1,000 letters most of which were anecdotes of precognition experiences. American Robert Nelson set up a Registry for Prophetic Dreams in New York City. From a list of about 8,000 dream accounts sent to him, 48 (0.6 percent) were judged to be strong evidence of precognition. Rather counterintuitively and dodging statistics, that is enough to establish a basis for treating the phenomenon seriously; http://dreamhawk.com/dream-encyclopedia/esp-in-dreams/. The point is that very few people are equipped to see events beyond the present, just as very few people can run a marathon in less than 2 hours and six minutes as a calibrated stop watch can prove.

2   This seems to succinctly underscore what I have shown in this book. The quote is from *The Reality of Nature* by journalist Andrew Zimmerman-Jones, published July 2008; http://www.pbs.org/wgbh/nova/blogs/physics/2015/07/are-we-living-in-a-computer-simulation/. My answer to the last question is that we are living in a copy of a computer-guided existence first actualized in the multiverse Template.

3   See   http://www.jamichon.nl/jam_writings/1988_guyau_idea.pdf, and   http://www.ibiblio.org/HTMLTexts/John_Alexander_Gunn/Modern_French_Philosophy/chapter2-3.html.

4   J. A. Gunn, *The problem of time: an historical and critical study* (UK: George Allen & Unwin, 1929).

5   F. Macar and V. Pouthas, "Time, Action and Cognition: Towards Bridging the Gap," in F. Macar, V. Pouthas and W. J. Friedman, eds., *Time, Action and Cognition* (Dordrecht, Netherlands: Kluwer, 1992).

6   J. W. Dunne, *An Experiment with Time* (London: Faber & Faber, 1927).

7   "There is nothing new under the sun"! In this regard, he is leaning toward Aristotle's view that time is "fundamentally linked to change and movement." See Wikipedia article on Aristotle.

8   *Intrusions?* published in 1955 is based on an unfinished manuscript rescued by his widow, Cecily, and her son. It was published by Faber and Faber six years after his death.

9   Ibid.

10 This would best correspond to the mental state that R. M. Bucke, MD, called "cosmic consciousness," which was the title of his book published originally in 1901 and reprinted in 1961 and 1989 by Citadel Press. If so, Dunne would have been the youngest known recipient of this state not at all corresponding to Bucke's assertion that this state comes to a person only at about age thirty.

11 E. T. Wooldridge, *History of the flying wing: Early flying wings (1870–1920)* (Washington, DC: Smithsonian Institute Press, 1984). The airplane he refers to corresponds to the D.8, which was test flown in 1912. Thus, the time lapse from dream to realization was twenty-four years. This is normally regarded as a rather long interval thus reducing the probability of its being a robust correlation, but the several corresponding details are so unique that it should be regarded as an acceptable precognition.

12 *Intrusions?* published in 1955 is based on an unfinished manuscript rescued by his widow, Cecily, and her son. It was published by Faber and Faber six years after his death.

13 Ibid.

14 *Macrozanonia macrocarpa.* Dunne's aircraft designs did not conform to the integrated delta form of that seed but only to the warp of its stubby wings.

15 http://www.digplanet.com/wiki/John_William_Dunne.

16 Termed "wash out" by aeronautical engineers.

17 T. Besterman, *Proceedings of the Society for Psychical Research* 41 (1933): 186–204. Dunne's score (23.5 percent) was far above average; see also appendix 4. Even if Dunne's score is halved say on the grounds that Besterman used the term *suggestive*, it would still be above average. Later, Dunne's 1927 protocol was shown to be effective in an experimental setting. I have confirmed that by experiencing a precognitive dream. See also Jackson, *American Society for Psychical Research* 61 (1967): 346–53.

18 There is extensive information on the Internet about our connectedness (type: "we are connected"), but some of it has the atmosphere of a crusade and implies that we have the potential to take advantage of our immersion within a gigantic *entanglement*, a term from quantum mechanics. This is totally beyond what can be deduced from the STZM.

A single example will suffice to show that contrary to some modern beliefs, entanglement cannot be taken as a way to our salvation. In the copy procedure, a slice of an event was equated with the Now moment (a snapshot). Follow the COPI routine and focus on just one complete event. I have selected a gruesome one. A hidden sniper pulls the trigger; the bullet travels sixty meters in well under half a second and enters the head of a platoon leader tasked with surprising an enemy encampment the sniper was protecting. The two men are momentarily connected by the bullet, but that did not do the dead soldier any good. Gruesome things have been happening to humans (and to other organic life) since time immemorial. Current conditions on this planet are simply an extension of the same thing.

19   J. W. Dunne, *An Experiment with Time* (London: Faber & Faber, 1927).

20   Ibid.

21   In late 1913, a well-known rift occurred between Freud and Jung over Jung's belief in paranormal phenomena. Jung and Pauli stayed together because they realized paranormal happenings are not an illusion but part of life.

22   This term is attributed to Jung. I have my own term: *serial significantly linked events* (SSLE). The more the events and the closer the events are in clock time, the more statistically significant is the SSLE. The Pauli effect may be called a synchronicity.

23   A. I. Miller, *Jung, Pauli and the Pursuit of a Scientific Obsession* (New York: W.W. Norton, 2010).

24   Actualized in *The Practical Use of Dream Analysis* published in 1934.

25   Prominent examples are Uri Geller and Ingo Swann; http://en.wikipedia.org/wiki/Ingo_Swann.

26   Remo F. Roth provided the diagram in an article entitled "Wolfgang Pauli and Parapsychology (Part I)": http://paulijungunusmundus.eu/syn-w/pauli_parapsychology_p1. The year of the dream is given as 1934.

27   On September 29, 1927, a review of Dunne's *An Experiment with Time* was published under the title *Dreaming of the Future* in the *Times Literary Supplement* (London). The reviewer, J. W. Sullivan, a mathematician, was probably the first to question (in print) the validity of Dunne's weird system of time in a seemingly infinite regress. R.L. Mégroz quickly responded with a letter to the *TLS* pointing out that Dunne's account of his precognitive dreams was not at all unusual and that he (Mégroz) was collecting anecdotal accounts for a dream anthology that included precognition.

28   R. H. Canary, *T. S. Eliot: The Poet and his Critics* (Chicago: American Library Association, 1982).

29   Richard Wollheim (1923–2003) was a British philosopher and professor of mind and logic at University College London; he wrote a book about the philosopher F. H. Bradley.

30   We learn that initially, Eliot was influenced by the ideas of Bergson but later became disenchanted with him. This is evident in the stanzas about time in "Burnt Norton." The reference to and the identity of Herr Smidt is unknown.

31   S. Bergsten, *Time and eternity: A study in the structure and symbolism of T. S. Eliot's Four Quartets* (Stockholm: Svenska Bokfürlaget, 1960).

CHAPTER 5

Summary and Discussion

It is often more difficult to deliver the hypothesis than to discover it.
—Modified from a quote by Jean-Baptiste de Lamarck (1744–1829) who was an early pre-Darwinian explorer of biological evolution.

Overview

This chapter expands on material in chapters 2 and 3 and collates enough additional evidence to make a convincing argument for the existence of a serial time-zoned multiverse. It has the potential for causing a general paradigm shift in current scientific and ontological thinking. Such a paradigm shift has already been in waiting and scattered in different places for the last many decades. For some, it may have already arrived. Though my original goal was much more modest, I continued to find sufficient dots of information to make a case for a purposeful multiverse; that is, purposeful for humans. Additional dots appeared while I was writing this chapter. Section headings here resemble those in chapter 3. Be prepared for the longest chapter in this book; it was mainly due to the Internet that this happened.

While we hear from a miniscule but by no means intellectually insignificant minority the assertion that our existence in this universe as well as the universe itself is pointless, this clearly hasn't deterred mainstream research in the physical, biological, and sociological scientific fields. Neither have the sterling efforts of minority groups investigating the preternatural (a.k.a. the paranormal) had any significant effect on mainstream scientific research. Thus, a curious, quasi-stable polarization in these research fields seems to have developed.

While population levels continually and monotonically climb, ethnic groups keep shuttling around the planet with the ultimate goal of some people is to transport human colonists off this planet to another one. There is not enough basic thinking in the areas of human ontology and little in the way of population management. The clues to our existence basically reside in three types of well-known phenomena: out-of-body experiences (OBEs), precognition (prescience), and (re)incarnation. Most people working in the mainstream sciences usually view these phenomena as grey areas and brush them aside. First, they are immediately seen as not amenable to physical analysis. Second, the data are mainly anecdotal and obviously not individually repeatable, so they are disqualified from publication in mainstream journals. Third, there are insufficient sources of funding. Thus, these areas of research are largely ignored.

In 1929, Arthur Eddington stated that to solve the mystery of our existence, we should be free to use spiritual and preternatural knowledge as well as established science. He importantly predicted that the key factor involved in this task would be time. His general attitude toward using all types of data is very similar to that of physicist Wolfgang Pauli, and there have been many recent additions to this list from those inside and those with one foot outside mainstream physics.

Over the last 130 years, the few well-known scientists who have risked their reputations by conducting experiments to prove or disprove phenomena such as telepathy, telekinesis, and even reincarnation to name a few show that the spark of radical inquiry is still alive. Yet the main question facing physical scientists and the experimental preternatural investigators is what the basic reason behind our existence is. Was the universe built for us? Or are we as judged by Stephen Hawking merely "chemical scum" that happened to develop in this seemingly hostile cosmic environment after a suitable atmosphere had formed on our planet and bombardment from meteorites had abated to statistically comfortable levels? The truth is that we are built of the elements formed in this hostile environment; we are part of it.

Recall Dunne's remark: that we must go along with the rest of the contents of the universe. He further asserted that no special treatment is afforded us. To this I say: *yes* and *no*. The *yes* applies only to our physical bodies. The *no* applies to our spiritual bodies which are afforded special treatment by way of immortality, and we all seem to have souls. They arrive in newborns by an unknown process, they suffer or enjoy lives along with the physical bodies, and they are able to exit the physical bodies by some unknown, automatic process. This applies in cases in which a serious threat to the safety of the body is sensed, temporarily (producing OBEs and NDEs) and permanently (as in death of the physical body).

This process has to be acknowledged if we want a complete understanding of our existence. We may not be able to do anything about this, but it should enable us to experience a more meaningful life and not to be persuaded by some that life is pointless. Knowing that our souls came here expressly for the experience and that they can safely exit the physical body may help us stabilize our minds.

On the other hand, I have argued elsewhere that it seems intended that we should live our lives as they were lived in the Template and not be sidetracked by illusions. At least, we have a choice, and we see that people do make the choice. I tend to toggle between the two states depending on the situation.

In the last half century, the increase in levels of scientific research and of the number of publications in established areas of research seems to be almost exponential in character. Research in areas of the preternatural is also climbing. As I was writing this book, I sensed that interest in and discussions about multiverses were increasing but that there were no obvious advances in knowledge about the structure of the multiverse. This aspect has turned out to be essential for understanding the nature of our existence and the puzzles surrounding us as revealed by a thorough interrogation of the serial time-zoned multiverse.

It would be a mistake to classify this book as fringe science fiction or think it rests entirely on metaphysics. By current standards, there may be metaphysics in here, but I argue that it is not detrimental to the STZM model, which is clearly verifiable in multiple ways. Other multiverse models also contain metaphysics; that indicates that this practice is now tolerated. Indeed, some of these models have been conceived by physicists holding tenure at major universities. They are apparently taking seriously Einstein's declaration "Imagination is more important than knowledge."

Perhaps I can add that logic can sometimes be more important than writing equations of formal mathematics that may contain errors caused by making dubious assumptions along the way. Pauli once said that no progress with trying to find a link between classical physics and the preternatural realm will succeed using logic alone. Geometry combined with logic seems to be what is needed to get results.

Bernard Haisch rightly argued the case for a purposeful universe. The way to recognize our universe as purposeful involves three essential elements: the first is to recognize that we are living in a specially configured multiverse; the second, just as important, is that it is essential to demystify the word *time*. The third element is to understand that we are essentially indestructible spirit beings of energy temporarily lodged (as souls) in physical bodies and that these souls are there for a specific purpose—to experience selected lives. This is basic spirit speak and known by people since antiquity. Therefore, not only is this a purposeful universe, but we are also automatically and more powerfully inside a purposeful multiverse in which there is evidently communication between individual universes. It can be said that precognition is no less than a window through which we can access the multiverse.

A wealth of precognition anecdotes span millennia and cover vast tracts of the planet's population. In this connection, how many cases of precognition or premonition does it take to be convinced that these phenomena exist? Is this a situation where one is allowed to say: why be satisfied by one case when millions or even billions will do? As it turns out, the sheer volume of data on precognition (and premonitions) before the turn of this millennium was a revelation to me, and it kept me googling for many days and nights.

If mainstream physicists object to what I have done here because it lacks mathematical rigor, it will have to remain that way. But I should draw their attention to books written by scientists who are arguing a perceived case built on some or all of the following actions: using combinations of approximations to ensure that the analysis remains within the bounds of mathematics, selecting experimental data, using thought experiments, geometry, and even statistics to convey their ideas. To illustrate this, consider now two well-known individuals who practiced entirely different approaches to doing successful science.

Nikola Tesla (1856–1943) and Albert Einstein (1879–1955)

Because of his seniority, I start with Tesla, a celebrated experimental electrical and mechanical engineer. Among many other inventions he is credited with designing is the alternator, which produces alternating current. He stated, "Today's scientists have substituted mathematics for experiments, and they wander off through equation after equation and eventually build a structure which has no relation to reality." He made this statement most likely in 1931.

On July 20 of that year, Tesla appeared on the cover of *Time* magazine as Man of the Year. Einstein would have taken note of that. In June 1931, he wrote from Potsdam a congratulatory letter to Tesla on his seventy-fifth birthday; that meant Einstein must have had considerable respect for Tesla. Indeed, one of Einstein's later philosophical quotes seems to run in a ghostly fashion parallel to what Tesla wrote above. Einstein stated, "As far as the mathematics is certain, they do not refer to reality." This is not what he said early in the successful application of his general relativity field equations to selected astrophysical/cosmological problems.

By 1931, Einstein had already established himself as a famous theoretical physicist mainly because of his development of the theory of general relativity, which purports to explain gravity as well as space and time in a radically new way. But bear in mind that he had substantial help from several others in achieving this goal. A media interviewer later asked him how it felt to be the smartest man alive. Einstein replied, "I do not know. You will have to ask Nikola Tesla." That remark had to have been made before January 7, 1943, when Tesla died. This remark appears especially odd because Albert was one of those who wrote "equation after equation." In fact, Tesla had made no secret of the fact that he was suspicious of Einstein's relativity theory though it was really none of his concern.

Nevertheless, even by 1931, several aspects of Einstein's new physics had been verified (by Einstein and several others) using experimental data. Very significantly though, the large-scale results —which could only be achieved by making approximations —were not cast in relativistic terms. These were Friedman's cosmological solutions which Einstein accepted only begrudgingly.

The approximate correspondence with reality is shown by the fact that the universe is expanding but not at the accelerating rate that current observations seem to indicate. His theory was clearly incomplete, a deficiency he later leveled against the quantum theory physicists.

In 2012, being an outsider to the highly charged atmosphere of scientific journals where it is important to be linked into their research community, I began to wonder where I could get the fledgling manuscript of this book published. I discovered the Society of Scientific Exploration and their publications the *Journal of Scientific Exploration* (JSE) and *Edge-Science*. These periodicals contain articles by respected, open-minded scientists who hold or who have held positions at major universities in the United States and elsewhere. Some articles appear to be on the fringe of physics; mainstream scientists would call them contentious or speculative, but one could say the same about the nine different theories of the multiverse that were then in circulation.

I duly submitted a manuscript to JSE on the multiverse synthesis covered in chapter 2. Over two months later, I made inquiries about its status. I discovered that certain members of the editorial board of JSE did not believe in multiverse theories. I received comments such as, "We have had enough of these multiverse theories already" and mockingly, "Perhaps there is another universe in which there is a journal that specializes in multiverse theories."

The Block Concept of Time Briefly Revisited

A Mental Block

By way of introduction, I will relate an anecdote about physicist Richard Feynman. He once avoided defining time for interviewer Damien Broderick by saying it was much too hard to explain. It did not directly concern him because his equations in quantum electrodynamics worked perfectly well without worrying about that. The same applied to Isaac Newton. That is also the case for Einstein's relativity equations but with a different spin. The truth is that defining time is not too difficult to handle if you block out all the confusing things that have ever been said or written about it. Even physicists can get themselves confused by others who went famously before them. Modern concepts such as the relativity of time don't help matters either. What one should be considering is the relativity of events.

In his venerable 1964 Lectures on Physics, Feynman wrote, "The test of all knowledge is experiment. Experiment is the sole judge of scientific truth." One has to assume, though, that the experimental results are correct. This is where the credibility of the serial time-zoned multiverse leans heavily on a wealth of experimental or empirical verification. This is where one has to face the resolution of some paradoxes. By solving them, even they in principle provide valuable tests. These data keep appearing. Then there are the data that may come later. It is a possibility that future quantitative measurements may be made to check out specific parts of the STZM model as put forward later in this chapter.

The Story of Vitalizing the Block

It was explained earlier that block time can be represented by a series of Now moments held in stasis (as in a thought experiment) and analogous to the laid-out filmstrip with its series of frames showing a progression of changes locked in the chemical emulsion bonded to a strip of celluloid. This incomplete concept must be taken a step further using the thought-experiment approach. The key question to ask is, Just where are all these Nows? It seemed obvious because of the assumptions made in chapter 2 that we can claim only one of them for this universe.

It became clear that the row of block Now moments implied the existence of a multiverse, each Now belonging to a separate universe. Furthermore, in reality, each universe has motion just as this one does. A continual series of Now moments pass through each universe via the COPI command. If you see the clock on the tower holding Big Ben in a movie, the hands are frozen in each frame along with everything else in that frame. That clock time registers a Now, or equivalently, a snapshot. When these serial Nows are viewed from outside the multiverse, they would seem to represent what physicists have been calling block time, but they should be regarded as a block of Now moments. These are represented by the dots along the temporarily locked horizontal dynamic events line (*DEL*) of the Blueprint.

Next step: allow all block Now moments to be unlocked so that motion resumes in each universe. This means that each one has its own time zone; the time in any one universe (except the Template) lags with respect to the universe in front of it. Each of the clock times belongs in a cosmic time, which is the same time that the particular Friedman-Einstein universe operated in. The resulting series of time zones is roughly analogous to our planetary time zones. Recall that lowercase *cosmic* is reserved for use within or just enclosing any one universe while uppercase *Cosmic* is reserved for the multiverse as a whole. This is where type-1 time operates. By hypothesis, it is manifest in individual universes as the frequency f of refreshing the Now moment. Thus in principle, it is made accessible to us. Einstein was at a loss as to how to do this; he said it was outside the laws of physics.

Though the physics term *cosmic time* seems to have been in use for only about a century, its proper use has not always been well defined or appreciated; this is due largely I believe to the modern concept of the relativity of time. The original concept of cosmic time,[1] or absolute time for our universe, predates Newton; it may even have roots that go back millennia though on less-substantial grounds.

Use of the two types of time: Cosmic (type-1 time) and cosmic (type-2 time) was found to be essential in the context of completing the Blueprint, which was developed from my immediate need to make any headway with the problem I was originally confronted with. I have just recently discovered that David Bohm argued in his 1993 book for the existence of an "absolute Now" based on his holistically viewed universe and his ontological interpretation of quantum physics.[2] This would evidently be referring to just our universe (U_8), but although he did not specifically consider a multiverse, the argument is trans-ferrable to each universe in a multiverse. The phenomenon of instan-taneous action at an indeterminate distance (known as nonlocality and linked to the term *entanglement*) could be active at all scales; this is something that is virtually required by the Δt-driven, cut-out and paste-in procedure operating between universes. Herein lies a mechanism I could not avoid addressing in terms of geometry, but it bears improving on by the quantum computer software aficionados. The mechanism of copying from one universe to the next requires replacing the current Now moment with a new Now moment that was our imminent future just a finite Δt moment ago in the universe ahead. Thus, the information in a universe (all phases of inorganics and organics) has to be transferred as instantly as possible in a single file! Herein was Bohm's vision—everything inside a universe must be considered as a whole, and every part is interconnected. This leads us inexorably back to Bohm.

Bohm's Model of Reality and his Concept of Time

Bohm and the New Physics

In the 1983 ARK edition of Bohm's *Wholeness and the Implicate Order*[3] (WIO), the editors provided a very helpful abstract only part of which is reproduced here.

> Wholeness and the Implicate Order proposes a new model of reality. Professor Bohm argues that if we are guided by a self-willed view we will perceive and experience the world as fragmented. Such a view is false, because it is based on our mistaking the content of our thought as a description of the world as it appears ... his concept of totality includes both matter and consciousness.

In Pauli's view, this would not go far enough. Human consciousness must be qualified by stating that through it, one can experience precognition, be able to see spirit forms, and be involved with synchronicities and SSLEs and take them seriously as Jung and Pauli did. If this cannot be achieved, I would posit that a person's consciousness is not fully developed.

I have omitted Bohm's ideas involving holographic applications because they are in the details category here and because I don't fully understand them –particularly the enfolding and unfolding processes as has been expressed by others. One reviewer (A. Shimony) judged it to be a work in progress. Basil Hiley[4] had to translate (meaning extensively edit) Bohm's last manuscript before it was posthumously published.

Bohm points to the pervading quantum matrix as being a field filling each universe. This appears to be crucial to his term *wholeness*. One must realize that the term *field* has to be treated not in the sense of a mathematical continuum but in the sense of a pervading, three-dimensional, gridded, digital form, a level I assume must be below the one in which quantum mechanics operates. Such a formulation in principle (I might naively say) allows objects existing in the field to be driven by a quantum computer. These statements constitute a provisional conjecture beneficial to my thinking but probably very little else.

One other concept about which Bohm was quite explicit was the space and time domain he was thinking in. He wrote, "Newtonian physics allows us to understand much of material reality ... as ... a special case." Here, *special case* refers to its relation to general relativity, which I have already avoided as being irrelevant to the theme of this book with the proviso that there is *event* relativity in the STZM, whereby it is associated with time. This is seen if one uses McTaggart's 1908 B series of time view. So again, Bohm was taking a basically Newtonian approach in his thinking about WIO just as I argued it was necessary to do before constructing the STZM model. This assumption was necessary to proceed as far as I did in chapter 2. I now refocus to this larger scale of our existence in order to analyze Bohm's concept of time.

Bohm[5] wrote that we were confronted with having to deal with

> a fundamentally new notion of the meaning of time ...
> It is derived from a <u>higher-dimensional ground</u>, as a
> particular order ... One can further say that many such
> particular interrelated time orders can be derived for
> different sets of sequences of moments <u>corresponding
> to material systems that travel at different speeds</u>. (emphasis added)

Does this description seem compatible with the serial time-zoned multiverse? Bohm seems to me to have been on the verge of assuming a multiverse. His invented terminology presents the only real challenge here. In what follows, Bohm's moment can be identified with the Now (I have already deliberately used the complete term: Now moment).

Just as Whitehead did, Bohm has forced me to pay closer attention to making clear statements about time. The time pips of duration Δt

that occur in the STZM model closely constrain the length of the Now moment, and equally important, they determine the timing of consecutive incremental positions of each universe as they travel forward at different speeds in the torus field. This is the view formed external to the multiverse. The next figure will geometrically illustrate the dynamics of this. If you are present and are thinking inside a universe, you will not be aware of an ultra-rapid flickering as the Now moments tick by. But that wouldn't prevent you from philosophizing on the nature of the Now moment as Einstein had done. I followed an external approach to the Now moment by conceiving a mechanism that advanced universes in their predetermined track simultaneously with receiving COPI files traveling in the reverse direction.

Bohm continued,

> However, these [material systems] are all dependent on a <u>multidimensional reality</u> that cannot be comprehended fully in terms of any time order or set of such orders … The fundamental law, then, is that of the immense <u>multidimensional ground</u> and the projections from this ground determine whatever time orders there may be. (emphasis added)

> One must then go on to a consideration of <u>time as a projection of multidimensional reality into a sequence of moments</u> (emphasis added)

In summary, here are the three main elements of what Bohm was saying as underlined and my brief comments on them as I attempt to relate his scheme to the STZM model.

1.   Time is derived from "a higher-dimensional ground," "a multidimensional reality," and an "immense multidimensional ground." The multiverse seems to fit at least one of these descriptions. Taking a dimensional reality to be a domain then the number of universes in the multiverse is equal to the number of domains.

2. He established a "time order" then wrote, "Many such particular interrelated time orders can be derived for different sets of sequences of moments." This seems to be related to the time-zone concept, seen in the STZM cosmological model in which the flow of discrete moments in a particular copied universe operates equivalently with the flow of moments around the multiverse. By considering "time as a projection of multidimensional reality into a sequence of moments," Bohm was anticipating exactly what I did in moving from figure 2.1A to figure 2.1B.

3. The statement "The moments correspond ... to material systems that travel at different speeds" fits exactly with what is seen in figure 2.1B, where the universes due to their changing sizes must all be moving at different speeds in the torus. Given some dimensions, mathematical equations could easily be written quantifying the relative speeds.

This concept of the Now was found to be fundamental in describing the steps in the early part of the multiverse model development. Here is how Paul Davies[6] described the conundrum: "Even Einstein confessed ... that the problem of the Now worried him seriously." He knew that there was "something essential about the Now" but reasoned it existed "just outside the realm of science." This might have come from the thinking shared by Bohm and Einstein, who were in close contact at Princeton University's Institute for Advanced Study up to about 1950.

Julian Barbour[7] also quoted from the same source[8] as Davies had regarding Einstein's bafflement over the nature of the Now. Einstein was directly recorded as having said, "The Now means something special for man, something essentially different from the past and the future, but this important difference does not and cannot occur within physics[9]." Based on the apparent compatibility between Bohm's statements and the structure of the STZM, I took up the challenge to investigate the possible dynamics of the Now in a geometrical frame-work. This is expressed in figure 5.1 in two related parts of vastly different scale. There is a direct coupling between the two diagrams but not an overall registration, which isn't possible to accomplish in such a diagram.

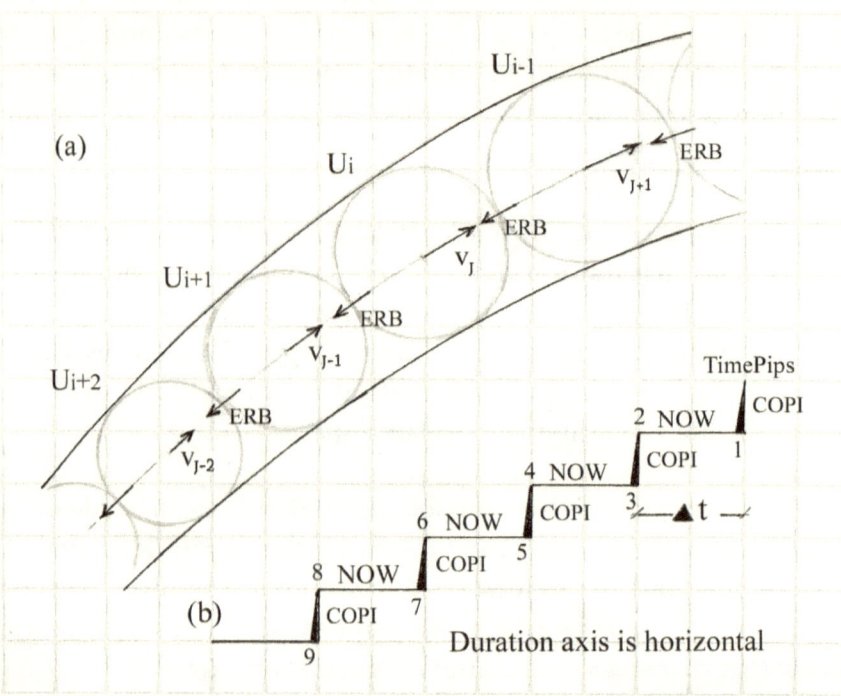

Fig 5.1(a) Space domain group of universes ($U_{i+/-subfix}$), within the 3-d multiverse torus bounded by the curved surface of the manifold (shown in section as lines). For i = 8 (our universe) the Template universe (U_t) is off the graph at far right. Universes expand as they move at speed $v_{j+/-subfix}$ where this j is just a clocked number, as in the j used in the $N_{i,j}$ notation. Universes move clockwise, doing so in small incremental steps, as illustrated in graphic (b) which, at such small scale, cannot be brought into registration with (a) as in reality it is supposed to do. The ERB label, explained in the text, is connected with information transfer located in the small gap between universes. There are many micro-steps of movement involved before U_i (for example) arrives at the position now occupied by U_{i-1}. These steps are shown in **(b)** which is a zoomed-in conceptual diagram within the 'process-duration' domain. The 'process' is a COPI procedure delivered through each ERB shown in (a). This 'process' is represented by the slender black triangles and is explained in the text. The Time Pips (TP) existing in duration, along the horizontal axis, are contained in Type 1 time. They are labeled at the apex of the black triangles with even numbers. The TP's drive the multiverse processes at a specified rate.

The black-filled Δt is the period of the oscillation. (The frequency is expressed as $1/\Delta t$). The duration of the 'Now' is either very close to Δt or equal to it. During a Now moment, each universe moves forward a small distance –not shown in (a) –which is determined by its position in the pre-existing, constraining field constituting the torus manifold. The sequence of steps: $1 \rightarrow 9$ are explained in the text.

Figure 5.1 conceptually helps explain the relationship between the too-small-to-be-depicted, micro-movements of a group of universes in diagram a and the Nows, the $\Delta t's$, and the COPI procedures in diagram b. The Nows are intentionally unnumbered as they are continually being replaced in each universe. Expressed in formal notation, all these Nows, say in a specific universe U_i, would be a set of N_i (written as $N_{i,j}$ where i is defined as the universe number and j is counting the individual Now moments as they pass through).

Universes move in the direction of the arrows that apply at the center of each sphere. The velocity vectors v_{j-2}, v_{j-1}, v_j, and v_{j+1} (where **j** is an integer that defines the universe and is not related to the j used earlier) show that in this section, universes speed up as they move forward and expand inside the torus manifold that contains the universes. This is not the case for universes situated beyond $\theta = 180°$ as seen in figure 2.1B.

According to the MIW multiverse model, finite gaps should occur between universes. This implies that like-signed electrical charges exist on the periphery of universes that make them slightly repulsive. This force is assumed to be countered by gravitational attraction. I now enter the metaphysics zone.

These gaps could be occupied by an Einstein-Rosen Bridge, labeled ERB. They are based on Einstein's gravity theory in the theory of general relativity. Physicist J. A. Wheeler got away with referring to them as wormholes. A stubby worm tube with bell-shaped ends would be suitable. Geometries of this type may be found on the Internet. This suggests that open-ended ERBs could be involved in the inter-universe information transfer system. (Here is an example of what I claimed was to be avoided: getting into the details). The reverse-pointed arrows through the ERBs indicate the flow direction of information, in this case, copied data. Next, refer to diagram b, which illustrates the steps in a multiverse-wide, synchronized process.

Start at the upper right of the diagram; run your eye along it from the top right to lower left in noncalibrated time expressed simply as duration. Begin at location 1 in the universes that had just prior to this received (as indicated by the arrow) the Now displayed between 1 and 2. From there, the next instruction would be to cut out and paste in (COPI) to segment 3–4. Alternatively, it could be a copy and paste in depending on the requirements of the programming mentioned earlier. In synchrony with this procedure, all the universes move forward a small step. This sequence of events is repeated along to location 9 and so on. This procedure means that any particular universe file contains an update of every particle and solid object in that universe.

The time pips (TP), with a frequency of $1/\Delta t$, can be considered the pacemaker of the multiverse. There is no reason why that oscillator couldn't also be used on the creator's side of the multiverse to drive a clock to measure the time since the GE. The Now segments in diagram b are aligned sub-horizontally here. They correspond to the vertical lines in figure 2.1A. Those scaffold lines include all the historical Now moments that occurred in the events that happened previously in that universe. Recall that only the most recently experienced Now in each universe is at the top of the triangular part of the diagram (in the *DEL),* as shown in figure 2.1A. Likewise, in figure 5.1b, only the current Nows can be displayed.

The Essential Structures of the Blueprint and the Cosmological Model

Two Aspects of the Same Concept

The Blueprint can be considered analogous to an electrical circuit diagram with which some electronic device such as a radio could be built. The Cosmological model would correspond to the radio wave receiver. The Blueprint is unique, but the Cosmological model derived from it is not the only possibility; it is merely the most purposeful one, the most interesting one, and an elegant one. In physics, an elegant solution is often considered with respect; *so it should be in metaphysics.*

The important elements of both aspects of the model are the necessity for a Template universe to have been generated and a succession of serially copied universes that make up the rest of the multiverse. The scheme is necessarily serial in space; that allows the automatic generation of time zones and a sequence of events that conform to the principle of cause and effect. That principle strictly applies in every universe. This is where precognition has caused some investigators of that phenomenon to question causality. But this is a problem in the relativity of perceived time; the STZM solves that problem.

Several variant Cosmological models may be built from the Blueprint, which requires only a few obvious adjustments to the evolutionary event paths. The basic feature of the selected model is that it is looped and therefore capable of recycling mass and energy. It is thus more efficient and therefore more interesting than runaway multiverse cases. The special feature is that the number of universes is finite. You will see in the next chapter that this feature is just one of many that are in accord with the described architecture of the MIW scheme containing a specified quantum mechanical field.[10]

A constant question that arises is, where does the power come from for driving the multiverse? As you have seen, this includes implementing programmed instructions such as the COPI procedure and the advancement of each universe around the circuit. One can easily say that it resides in the matrix field, below the "quantum foam" as John Wheeler called it. This posits a background state of vibrations, thus energy. But the big teleological question is, What energizes the vibrations?

It's possible that many additional discrete fields share the same space with fields that our atmosphere contains such as the geomagnetic and gravity fields. An analogy might be to multiplexing used in information transmission.[11] To mobilize this transmission, there is only one concept that seems to provide a mechanism—the concept of a Cosmic quantum computer (CQC). As I currently understand it, the construction and the programming principles of an earthbound QC are under development at several locations.[12] This indicates the serious attitude taken by some scientists toward extending the envelope of consciousness.

A Brief Encapsulation of the Two Types of Time

My concept of time and how to express it sharpened as I wrote this book. It resulted in the rewriting of many sections based on an earlier, incomplete understanding of time. The Blueprint facilitated the initial identification of type-1 time (identified with Δt), and the Cosmological diagram allowed a possible physical manifestation of how it would operate. I subsequently identified this time as probably being what Whitehead called nonformal time.[13] As a process philosopher, he seemed to think through this concept of time purely in his mind. Type-1 time is not amenable to subordination. It may be assigned an absolute status, but its implementation is relative at least for the purposes of discussing the attributes of the proposed Cosmological model.

The following summary was understood by many people even long before I started the multiverse project. Type-2 time is the same as the formal time of Whitehead. This is our clock time and can easily be subordinated because of its amenability to being directly associated with events and thus motion, which was the reason for its invention in the first place. Examples of subordination were provided in chapters 3 and 4 using ancient and modern literature. The best way to give type-2 time a visual manifestation and to show its astronomical underpinnings, is via the time court (figure 5.2), a sundial with no mechanically moving parts. The shadow cast by the gnomon is the only indication of movement, and that of course provides the astronomical link. At the large scale, solstices and equinoxes are marked; at the small scale, the local hour of the day (standard time and DST) is indicated.

Fig 5.2 The 'Kiwi' Time Court in Hamilton Gardens, New Zealand. Due to lighting conditions only the start of the shadow may be seen at the base of the gnomon (vertical pointed post). It is a graphic way of grasping what is behind *type 2* or Whitehead's *formal* time by representing it in a geometrical framework, in which the position of astronomically determined events are identified by a pointed shadow contacting a labelled marker on the court.

At first, it looked as if these two types of time were independent of each other and that type-1 time was completely inaccessible, but when the go of the Now moment was interrogated by using the STZM model, it showed that type-1 time was manifested into the domain of type-2 time. It therefore became apparent that in principle, type-1 time could be interrogated by making measurements —but only in the type-2 time system. The measurement needed is a diagnostic frequency signal. Hey, presto. This would then leads to a value of Δt in our own units! But is it possible?

The Precognitive Dream: Sampling events in a Future Time Zone

The STZM model was claimed in principle to be able to account for the occurrence of dreams about future events in broad terms only. This is because there are by hypothesis as well as experimental evidence universes containing our future in which events are being played out (in accordance with the postulated state of eternality, which comes from the philosophical term of eternalism) and because of a COPI procedure that transports Now files and actualizes those events in our own space and time domain—this universe.

What is missing in this scenario is the mechanism whereby the images of the future events arrived in our dreaming minds (or the lower-level subconscious). One mechanism is where our consciousness becomes connected to and locked onto a specific universe situated far ahead of us in another time zone. It seems likely that the particular universe locked onto at a particular time possesses a significant part of the event that was transmitted. Some dreams look as if they had an early section of a target event, others the middle, and others the tail end —as in Lincoln's dream described in the section ahead: La Forza del Destino. If we knew the value of Δt and we recorded the duration between the date and time of the dream and the date and time of the actualization, it is a straightforward calculation to obtain the number of universes that must have been ahead of us when we had the dream.

Obviously, the Δt value would have to be small enough so our visual system would not pick up judder from the passage of the Now moments. Here, I am doing only thought experiments. If Δt was set at 0.01 (one one-hundredth of a second), such a low-frequency signal (100 hertz) might be interfered with by other spurious signals near that frequency, but it would be well clear of the very exact 50 and 60 hertz frequency spikes due to the presence of alternating current sources. At 100 hertz and if the duration between the dream and the actualization was one week, the number of universes traversed by the Now slices would be only about 60 million.

Another possibility is that it is a medium-frequency (megahertz) wave. There seems no alternative but to hunt for an unaccounted-for frequency from which to derive a tentative Δt. It would seem to be left up to a decisive measurement to be made on some suitable organic system (such as a human with a normal brain for EEG monitoring) or on an inorganic chemical system in an established growth phase. I will leave this topic and return to the consideration of prescience.

The most commonly reported accounts of seeing a future event are by far through dreams, but they also arrive as daytime visions. Thought to be basically holographic, some are static while others are dynamic. In preparation for the next section, the following is a very brief discussion of precognitive dreams. It seems possible that a certain power level may be ascribed to these dreams. At the upper end of the scale are the vivid nightmares. The energy level seems to be increased when loss of life is involved or anticipated, so there is a very subjective human assessment process involved here.

At the lower end of the scale, there may be only a seemingly innocuous series of scenes presented when in April 2013 I had a precognitive dream of that type. It occurred after nearly three years of logging dreams of interest. It contained some elements of symbolism in it as I recognized when it actualized in June of that year. This is typical of many of my dreams and indicates that some original information has been transformed or rendered into a more compact visual format possibly due to limitations in ways of sending information but also to make the information amount limited.

Then there are precognitions that have occurred under monitored laboratory conditions. The success of these experiments depends on the subject's level of psychic skill. One three-year study at Stanford Research International in the 1970s documented cases of remote viewing intended to be made in the present—in real time. However, it happened to include one spontaneous precognitive reading involving a Russian military target that was later checked out to be correct. This experiment was done by the number-one remote viewer in the United States at the time. I consider that the extensive evidence for precognition is not merely persuasive (as some overcautious journalists or commentators may state) but is an established, empirical fact. There is an overwhelming body of historical literature to support this claim.

Some well-known examples are laid out in the next section, and they all support predetermination in the individuals involved.

La Forza del Destino

Here, I present six well-documented accounts exhibiting the force of destiny.[14] The first one is so well known that I give only a short outline of the main details; the second comes from Samuel Clemens—Mark Twain. The third has already been mentioned in connection with the *Titanic* effect. The enormous loss of life involved prompted headlines in British newspapers in April 1912 providing documentation. The fourth one (9/11) is so well known that it requires only a short entry. The last two are downsized compared with the others because they were local events, more recent, and also well documented on the Internet.

Abraham Lincoln's Assassination Dream

President Lincoln was known to possess psychic tendencies. These took the form of precognitive dreams. For instance, in *Man and Time*, J. B. Priestley (see appendix 4) mentioned one that occurred before this one. It concerned the dream of the surrender of Confederate general Joseph E. Johnson at the close of the Civil War, which served as the prologue for Lincoln's assassination. I focus here only on the well-known assassination dream.

Lincoln experienced his foreboding dream[15] about two weeks (elsewhere 10 days or a few days) before he was fatally shot by John Wilkes Booth at Ford's Theatre in Washington, DC, on April 14, 1865. By April 11, Lincoln had become so agitated and annoyed about the dream that he related it to a number of people including Ward Hill Lamon, a close friend, confidant, and self-appointed bodyguard who recorded the dream in his diary. Lamon's daughter took possession of her father's diaries and they formed the basis for a book on her father's reminisces of Lincoln. It was published in 1895. There were also earlier newspaper articles that contained the story of the dream.

In the dream, Lincoln left his bed to investigate the source of sounds of subdued sobbing. After checking several rooms, he entered the East Room and saw a "sickening scene." A catafalque there was being guarded by soldiers. Mourners thronged the casket, which contained a shrouded body. The nonphysical form of Lincoln asked one of the soldiers who it was. The answer was, "The president. He was killed by an assassin." He was then woken up by a "burst of grief from the crowd." The full story is well known U.S. history but there are people who are bent on discrediting others who experience preternatural phenomena. Even Lincoln allegedly stated that he was skeptical about dreams and did not think he was the president who had died of an assassination. But he would not want to think that it was him in the catafalque. The third account in this section shows that W.T. Stead also did not want to believe his premonitions, but then why not forget them and stop relating them to others? This is an area for psychologists to study.

It is significant to this section that despite warnings from several people—Lamon, William Crook, who was an official bodyguard, and Secretary of War Edwin Stanton—Lincoln did not waver from his determination to attend the play on April 14. General Ulysses Grant and his wife, Julia, canceled their invitation to attend the play because the general's wife continued to have premonitions of some impending event that was worrying her. They left town that day for home. But there is no indication that they relayed their concern to the president. Also, incredibly, there is no indication of Crook's whereabouts that fateful night. A policeman (acting as a stand-in bodyguard) guarded the door to the theater balcony but was absent from his post (with no backup) during intermission. That was when assassin Booth, with his concealed Derringer, broke into the balcony. All these goings-on would have occurred in the Template universe in which forebodings and precognitions cannot occur. So to me, this drama speaks powerfully to *la forza del destino*.

Two weeks after the assassination, Lincoln's "casket was placed on a platform in the East Room of the White House, where it was flanked by soldiers—as in his dream." Of course, Lamon and others who had known about the dream may have been influenced enough to promote following the dream script down to the catafalque in the East Room with its guards. But the fact remains that Lincoln made his assassination dream known to several people well ahead of the actual event though the dream covered only the aftermath in its latter stages and was confined to the White House.

A few more technical details are in order here. Communication in dreams is evidently by thought transmission. I do not hear spoken words in my dreams but just the awareness of what someone is silently projecting into my mind. So the soldier received the question in his mind. As soon as the answer was formed in the soldier's mind, the out-of-body soul consciousness of Lincoln received the answer. It is often convincingly claimed that such communication can actually be accomplished between two or more persons in our space reality; it's known as telepathy. This ability was demonstrated in the presence of some notable people: Einstein, Freud (a disbeliever in such matters at the time), and Joseph Stalin not to mention many well-known others.

Samuel Clemens—Mark Twain

When Sam Clemens was about twenty-three and his brother Henry was nineteen or twenty, they were crewmembers of the paddle steamboat *Pennsylvania* that operated on the Mississippi. Sam was the steersman and Henry was the mud clerk. It was either late May or early June 1858 when the boat was docked in St. Louis. One night when Sam was overnighting at his sister's house, he had a shockingly realistic dream, one we would now call a lucid dream. The scene was his sister's sitting room where a metal coffin rested on two chairs. In it he saw Henry. On his upper body was a spray of white flowers topped with one crimson flower.

Upon waking early the next morning, Sam was overcome with grief thinking that this had already happened and that Henry was laid out in the living room. He went outside for some fresh air deliberately bypassing the living room. After walking half a block, he assured himself he had been dreaming, so he returned to look in the sitting room. There was nothing out of place there. He told his sister about the dream. They passed it off as just a dream.

The *Pennsylvania* sailed for New Orleans, and Sam had his mind taken off the dream. Besides, as steersman, he had been having serious disagreements with the regular pilot; the captain decided to get a new pilot or transfer Sam to another boat when they docked in New Orleans the next day. What actually happened was that Sam transferred to another boat, the *A. T. Lacey*, which was due to sail on the return trip up the river two days after the *Pennsylvania*. Henry stayed aboard that boat.

On June 13, when the *Pennsylvania* was just out of Memphis, the boat's boilers exploded killing 250 people. Henry was blown into the water badly burned, and despite trying to help save survivors, he quickly became incapacitated himself and was transported to Memphis for treatment. When the *A. T. Lacey* reached Memphis, Sam went to his brother's bedside and tended to him for several days. He asked for more morphine to subdue Henry's pain.

On June 20, Henry passed away. When Sam went to see him the next morning, he saw Henry lying in a metal coffin supported by two chairs. There was a bouquet of white flowers on his chest. Moments later, a woman entered the room and placed one crimson flower in the center of the bouquet.[16]

W. T. Stead, Passenger on the RMS *Titanic*

This is perhaps the most convincing of all the stories emanating from the saga of the sinking of the *Titanic*, and I have spent some considerable time (duration) studying this man from multiple sources available on the Internet which contains multiple accounts about Stead. This more than makes up for critics who still question Lincoln's dream account based on inconsistencies that are details and not giving sufficient attention to the main framework of the account.

From 1883 to 1889, Stead was the editor of the London-based *Pall Mall Gazette*, and he established his reputation as the father of investigative journalism. Some investigations were done in undercover style. His articles were sensational and blew a hole in the "stuffy atmosphere of Victorian journalism." He was a compulsive whistle-blower, muckraker, social and political reformer, and champion of the suppressed and the oppressed. One article that is pertinent here appeared in 1886. It concerned the dangers to shipping in the North Atlantic due to collision with other ships and the loss of lives due to an inadequate number of lifeboats. The *Gazette* folded, but in 1890, he managed to establish another tabloid, *Review of Reviews,* to continue his opinionated but certainly justified tirades against many controversial topics of the day.

In 1892, Stead published in his *Review* the fictional story of the rescue of a survivor (or survivors in another account) of a ship that sank after hitting an iceberg. The actual White Star Liner *Majestic* took the role of the rescue ship. Three years later, the captain of that ship was Edward Smith, who in 1912 was selected to captain the *Titanic*, which was built to replace the aging *Majestic*.

Already, it was becoming clear that Stead had another risky interest that was about to surface. In 1893, he founded a spiritualist quarterly called *Borderland.* According to one commentator, it was Stead's involvement in spiritualism that weakened his influence over his readers. The Society of Psychical Research had been established in London a decade earlier, and the issue of the paranormal was being laid out in the open there as well as in France and America. This was a time of great technological and scientific development, and the public was not prepared to spend time on weird, unexplainable stories. Two distinct reactions to the paranormal appeared: unshakable belief in it or the conviction that it was bunk, and ugly confrontations sometimes occurred.

I refer you to appendix 4 for more reading on the topic of precognitive dreams as of 1963 in England. The subject of such dreams has assumed rightly or wrongly one aspect of a vast field of the preternatural. The following information is taken from what Ian Stevenson, MD, a clinical investigator of past-life recall cases, wrote in the 1974 book from which information is collated and listed in a table with references.[17]

In 1898, Morgan Robertson's novella of the *Titan* was published[18]. Being a seaman and an amateur writer, Robertson was probably already familiar with what Stead had published about breaches of safety in the North Atlantic shipping lanes. Thus, even before the turn of the century, there were ominous warnings on the horizon typically directed at the threat presented by icebergs in the numerous shipping lanes.

Stead was known to have had hazy visions pertaining to his demise. Paraphrasing his words, it would involve either crowds or water. It turned out to be both; he just didn't think of putting them together because his dreams were just vignettes of the whole drama. His individual dream view shed must have been very narrow. He even gave some lectures in which he described some of his psychic experiences. There seemed to be an increasing concern that constantly worried Stead about the end of the first decade of the twentieth century. Then in early 1911, Count Harmon, whom Stead used to occasionally consult, told him that danger to his life would be caused by water. In June, Harmon wrote to Stead that travel would be dangerous for him in April 1912.

Over "some months" beginning in September 1911, Stead consulted another psychic, a sensitive named W. de Kerlor, who predicted that Stead would go to America, which at the time was news to Stead. That Stead did not reach America is obvious. Bearing in mind that psychics are not always 100% accurate must be taken into account. Kerlor might have said that Stead would be going or travelling to America; perhaps even that Stead would be leaving for America in April. He got that right and the year. Kerlor also saw "the picture of a huge black ship," the stern of which bore a "wreath of immortelles." Subsequent to this, de Kerlor had a dream in which he actively participated but which he said applied to Stead: "I was in the midst of a catastrophe on the water; there were … more than a thousand … bodies struggling in the water." After all this, Stead could only say, "Oh yes; well, well, you are a very gloomy prophet" as if to say, "I've heard this sort of thing before and nothing came of it." Recall what Lincoln was alleged to have said. It is a case of surficial denial.

It was because of Stead's literary influence that William Howard Taft, then president of the United States, had invited him to take part in a peace congress at Carnegie Hall. Though he knew he would be sailing across the Atlantic at a potentially dangerous time, Stead accepted this invitation despite all the warning signs—his own vague ones and those of two practicing psychics he had deliberately consulted. These last ones were persistent in their readings, and one offered more information that Stead hadn't solicited.

Survivors of the *Titanic* tragedy reported that Stead helped people into lifeboats and gave his life jacket to another passenger. At one stage, he was observed on the deck in a state of prayer as if preparing himself for the end. He probably didn't know that he had been nominated for the Nobel Peace Prize several times and that soon his number might be coming up. Instead, another number was rapidly coming up—it was a watery grave for him as it was for almost 1,500 others. While weakly clinging to a lifeboat and treading water, he might have been thinking, *Why did I not take those warnings seriously? Why did I believe this ship was unsinkable?* You might know the answer.

At the W. T. Stead memorial session on April, 22, 1912, at the Men and Religion Forward Movement meeting in Carnegie Hall, "Dr Newell Dwight Hillis said that on his last visit to America (1907) Mr Stead had told him that he expected to be stoned to death by a mob or otherwise to die by violence." He called Stead "the greatest of Anglo-Saxon Journalists."[19]

Here is a famous and intelligent man who communicated to the British nation and even to the Americans through his tabloids that there were extreme dangers in the shipping practices in the North Atlantic. He subsequently found himself a victim of those very malpractices. Is this a coincidence? I don't think so. Do you think as I do that this qualifies as the best recorded case of *la forza del destino* on record?

The Infamous 9/11

The Boundary Institute has listed 43 records[20] of premonitions and precognitions relating to the events of September 11, 2001. One notable record came from a CIA remote viewer (probably the one called number one) who in 1986 described what in hindsight matched the 9/11 event in close detail. This date was the earliest of the presages; most others were months, weeks, and even days before the event. Some dreams were symbolic and could not be deciphered intelligently enough at the time. The research of Rosemary Ellen Guiley[21] indicated that there were probably thousands of people involved in presaging 9/11 in some form or another, and she queried why nothing was done or could be done to prepare for it.

The Aberfan (Wales) Coal Slide

Larry Dossey[22] provided a concise, readily accessible account of the event that took place in Aberfan, Wales, on October 21, 1966. R. E. Guiley also referred to that disaster. "At least 200 persons" later reported having received psychic information that looked strongly like warnings of an impending disaster. Even a girl student at the school that was buried told her mother the day before that she had dreamed of a scene that later matched the description of the coal slide. The mother had no time to listen to the dream and so lost her daughter. One hundred and forty-four persons, mostly children, lost their lives.

It would seem that such a high concentration of warnings (which these premonitions and precognitions appear to have been) occurring in this case in a small, close-knit coal mining village should have been more than enough to cause a safety inspection to be made. But no action was taken. Guiley[23] wrote on her website in 2014, "Many people ask: *What is the point of precognition if it can't be used to change the course of events?*" This is the type of question that can be answered given knowledge of how a serial time-zoned multiverse model can give information on our existence. It is quite explicit about the principle of determinism in copied universes.

In contrast, Dossey presented several anecdotal cases in which it was asserted that because of prior information received, decisions were made that resulted in avoidance of injury or even saved lives. Thus, there is a questioning of data analysis. The trouble is we do not know how accurate these anecdotal accounts are. In Dossey's examples, these could have been cases of soft free will (as in the Dunne effect) in which energy levels played a role. On the other hand, in the introduction to his book, Dossey wrote that when in his first year of medical practice he "experienced a week of having premonitions about patients, all of which came true."[24] This seems to speak directly to the case for determinism. He also indicated that this ability lasted only a week and implied that he did not influence any of the outcomes. I have encountered and mentioned before the willy-nilly occurrence of certain preternatural phenomena. That is a study in itself but far beyond the scope of this book.

The main problem is being able to accurately record the dream before it recedes to a memory location that has so far not been identified. Precognitive dreams may contain elements of the past as well as the future making interpretation difficult. Because of the erratic and frequently clip-like occurrence with occasional symbolism added, precognitive dreams are almost never totally reliable as unambiguous warnings. Moreover, many people have a denial attitude towards dreams that contain a foreboding as in Lincoln's dream.

A Fatal Accident on Pyramid Peak, Colorado

This is a clear-cut case in which destiny ruled at a personal level. Dr. Heinz Pagels[25] was a physicist well versed in quantum theory and chief executive officer of the New York Academy of Sciences. He also held other important positions. His book *The Cosmic Code*, published in 1982, is a layperson's introduction to quantum physics and its relationship to our existence, so it is somewhat philosophical. In the last chapter, he drifted into his personal life using as an example his favorite sport of mountain climbing and his dreams about it.

Pagels told an older friend "I used to climb mountains covered in snow and ice, hanging onto the sides of great rocks[26]." The friend asked him why he wanted to kill himself. Pagels protested to that saying there were "rewards" such as "sight, pleasure and the thrill of pitting my body and my skills against nature." His friend countered with, "When you are as old as I am you will see that you are trying to kill yourself."

Pagels filled out some of the details of his climbing forays. He wrote that he often dreamed of falling.

> Such dreams are commonplace to the ambitious or those who climb mountains. Lately I dreamt I was clutching at the face of a rock but it would not hold. Gravel gave way. I grasped for a shrub, but it pulled loose, and in cold terror I fell into the abyss. Suddenly I realized that my fall was relative; there was no bottom and no end. A feeling of pleasure overcame me. I realized that what I embody, the principle of life, cannot be destroyed. It is written into the cosmic code, the order of the universe ... As I continued to fall in the dark void, embraced by the vault of the heavens, I sang to the beauty of the stars and made my peace with the darkness.[27]

It is clear that Pagels experienced a rarely reported out-of-body experience in a dream.

He spent summers at the Aspen Center for Physics in Aspen,

Colorado, a great getaway from New York, where he spent most of his time. On July 23, 1988, he was descending Pyramid Peak (14,000 feet) after making his seventh ascent of that mountain. His recently graduated PhD student Seth Lloyd accompanied him. Pagels had planted the seed for designing a quantum computer, and he had captured Lloyd's attention. This subject links to a later section in this chapter in which Lloyd and others continued the quantum computer story for us.

Carefully negotiating the treacherously loose rocks that sheathed the whole mountain was a very hard part of the climb and even more difficult on the descent. Paul Ryan[28] described the terrain as "consisting only of stacked rocks resembling dinner plates, platters, trays, tables, and even whole dining rooms, just waiting for some incautious hiker to send the whole mess down the mountain." Not just incautious hikers either. The loose rocks on which Pagels was descending started to slide uncontrollably. They quickly took him to his death. The shrub of his dream over seven years earlier might have been a figment of his imagination or a shred of memory from a slip on a climb somewhere else on another occasion, but the essence of the recurring dreams he had had was evidently a message he would be killed on a mountain.

I have a collection of memories that tell of people, especially mountaineers, who led dangerous lives and who took great risks sometimes despite having young families. Instead, they died in their pursuit of some community- to national-scale duty or for some personal-development needs. Such tragedies took place even after they had received psychic messages from other people and even premonitions of their own.

I suggest that skeptical readers present six cases of the same fidelity as the cases presented above that support the existence of free will to enable the TE effect. They need not be rated by the SPR or the ASPR. If a deadlock occurs, it's back to the drawing board. But perhaps this is a message to cease and desist? The next section is an outline and discussion of a conceivable computing and information transfer scheme in the STZM model. This could be outdated by the time the book is printed.

Outline of a Grand Scheme: the Cosmic Computer

General Comments

The STZM Cosmological model shares many of its descriptive characteristics given for the MIW model as shown in essay number 4 in the next chapter. But there are some conceptual snags I cannot understand. This may be due to the reports of commentators and their (as well as my own) inability to understand the math. The commentators obtained interviews with two of the authors. In my case, I sent one of the coauthors an outline of the STZM model and a set of questions, but I haven't yet received a reply before sending this manuscript to the Archway editor. This is not a good sign. Nevertheless, following are some of the problems I haven't been able to resolve.

In one commentator's interview, Michael Hall, the lead author, was reported as stating,

> The beauty of our approach is that if there is just one world our theory reduces to Newtonian mechanics, while if there is a gigantic number of worlds, it reproduces quantum mechanics ... In between it predicts something new that is neither Newton's theory nor quantum theory.
>
> —http://phys.org/news/2014-10-interacting-world s-theory-scientists-interaction.html#jCp

Because I am interested only in the multiverse case—involving a gigantic number of worlds—I am seeking to understand how the authors managed to arrive at a multiverse with specific properties that match many of the large scale properties of the mature STZM, derived by a different methodology, independent of quantum mechanics (QM). If their theory shows that for a single universe Newtonian mechanics is exhibited, is this only at the macro and larger scales? What then for QM? Does that have a Newtonian-like *time* existing contemporaneously in a single universe? Thus there is a puzzle here that I cannot resolve.

Einstein considered only one universe while constructing his theory of general relativity in which he concluded that Newton's theory needed to be generalized (hence modified) before several large space-scale phenomena could be explained accurately. Of course, quantum mechanics is absent in Einstein's cosmology theory, which occurs at a far different scale. To my knowledge, the two theories as they are now interpreted stand alone. The goal is to unify them, but it cannot be done. The devil is in the mathematics

The STZM theory doesn't feature quantum theory directly, but I showed how it is possible to discuss quantum processes within the STZM in terms of a quantum computer—one obvious way of driving the multiverse. The outcome of the STZM model is that it doesn't matter how many universes are in the multiverse as the torus is filling up; they all operate according to the same (neo-Newtonian) physics whether large scale or mesoscale. In the quote above the interviewer is writing that Hall is saying that the MIW theory predicts that the mechanics is changing as the multiverse is filling up. The STZM hypothesis is inherently short on details but it is hard to follow why the mechanics (of any kind) would change as events evolved. For example, there is no formal derivation that universes are separated by a small gap that MIW theory requires (or predicts), but it turns out that it is useful to the STZM model and thus conforms to common sense. So there are indications that all theories have their strengths, weaknesses, and supporting roles.

Thus, while the STZM model shares six of its properties with the MIW model as stated by its authors, I haven't been able to find out how this striking result happened. The conundrum is that the STZM does not have an explicit quantum mechanics scheme whereas the MIW does. Were these model multiverse properties reasonable assumptions that didn't interfere with the QM scheme or is there a physical coupling? We shall have to wait for feed-back on this book. Continuing at the quantum level we now shift back to the Cosmic quantum computer.

The computability properties of a multiverse or part of one are what David Deutsch[29] has been working on for over three decades. He specializes in the theory of programming of quantum computers. It is quite another matter to build a quantum computer. Quantum mechanical engineers have specified some of the components needed for a quantum-based machine such as for example various types of gates that act to route and control the information flow, but what of the real Cosmic quantum computer? In Deutsch's book was mention of computer scientist Tommaso Toffoli (then at MIT, now at Boston University) quoting him as saying that we are taking "a ride on the Great Computation that is going on already." Julian Brown[30] reported a later version of this that appeared in one of Toffoli research papers: "Nature has been continually computing the next state of the universe for billions of years. All we have to do is hitch a ride on this huge ongoing computation and try to discover which parts of it happen to go near to where we want." The statement to carefully watch here is to "hitch a ride…"

These types of almost metaphorical statements are the only ones I can bring in to my own naïve statements and count them in as perceived reinforcements to what I have written in this book. However, it is necessary to make clear that the Cosmic quantum computer's key role in the STZM model is to control the functioning of the whole multiverse. In contrast, Deutsch's multiverse is composed of the universes that result from the MWI of quantum mechanics. That is a very different multiverse, and there is no model for its physical structure or a provision for the universes to interact, which is a crucial requirement for the STZM and the MIW models.

Moreover, Deutsch assumes that a cosmic quantum computer exists in each universe and that the universes only "weakly interact" whereas in the STZM and MIW models they strongly interact. Another important point is: one cannot assume that 10^{500} universes are at your disposal if your multiverse isn't based on string theory, which is a hypothetical scheme dependent on mathematical manipulations and doesn't relate to the STZM or any other multiverse scheme.

A Red Flag

Extending these ideas, whatever the configuration is, the computer system is already tasked to perform operations essential to the operation of the multiverse. Is there enough capacity left for additional operations? They are easier to do, but they take longer with a serial multiverse; however, the small value of Δt is an important consideration. In any case, I shall naively raise the first red flag.

The serial multiverse arrangement seems to be implied by the situation that Toffoli was describing: that the computer, programmed to obey the laws of physics, is "continually computing the next state." This function is fulfilled in the STZM model, in which such a computation is hypothesized to occur in the Template, which I have already demonstrated is not the one we are living in. Note Toffoli's correct use here of *continually* rather than *continuously*.

The story Julian Brown[31] wove is far more informative, and many of my so-called technical questions have already been tackled there, but there is room for more questions. The idea that the universe is driven by a quantum computer also occurred to Edward Fredkin, who invited Toffoli to join him at MIT in the early days of quantum computer development. Fredkin was also in the forefront of designing gates for experimental quantum computers. Some of the present activity is directed toward local quantum computing that Toffoli achieved by using a plug-in on a conventional computer. A workable quantum computer is being developed independently by Seth Lloyd,[32] the former student of the late Heinz Pagels, who died in the foreseen mountaineering accident. Lloyd's target is to use small-scale quantum computing to understand our universe. Contrast this with Deutsch's ambitious plan to prove the existence of the multiverse.

The Other Red Flag

This second threat, which is associated with space travel, is of quite a different nature. A common factor in these situations is or seems to be the so-called zero point energy field (ZPF), which has been referred to as the matrix. The following is speculation generated by extrapolating from an idea that has developed in the last few decades.

The scheme of expending energy contained in the ZPF as a power source for propelling spacecraft on deep-space missions gained the attention of Harold Puthoff, the laser physicist, and one time researcher into remote viewing. Calculations by John Wheeler indicated that the ZPF contains in theory an immense amount of energy. The problem, they say, is how to engineer it. This situation may be less severe than was earlier thought to be the case as it has since been claimed that the amount of energy stored in the ZPF is much less than was earlier calculated. Imagine, however, the consequences if not one but thousands of rocket craft were traveling at nose-flattening speed throughout the cosmos vacuuming up energy and creating disequilibrium in the ZPF. Thus the pole of the second red flag is raised at the ready to 45° just in case there is a need to waggle it.

The following must be projected into the Template. The above situation brings to mind an idea I had in connection with our continual activities that have caused deleterious changes in the environment—causing dangerous modifications to the atmosphere and everything beneath it. This prominently includes the detonation of nuclear weapons in the atmosphere and underground. Is it possible we were deliberately positioned nearly midway between the infinitesimally small and the unimaginably large scales of the universe such that the effects resulting from our hyperactive minds and tinkering hands are minimized from doing major damage to the machinery of the universe? This issue of course is extended to the multiverse.

Some Tales about Physicists

Attitudes can build up in scientists' minds; here is an example: David Bohm sent a copy of his manuscript for the 1951 book *Quantum Theory* to Niels Bohr for his opinion. Bohr was known to have said about quantum mechanics, "If you are not shocked by it, you have not understood it." Bohm's manuscript included the idea of the hidden

variables attributed to Einstein, who could not accept Bohr's interpretation of the collapse of the wave function particularly when the collapse was believed to occur if it is looked at—however that is possible.
Bohr never replied to Bohm's letter accompanying the manuscript.
Perhaps the manuscript was not shocking enough!

This is a good place to bring in Wolfgang Pauli, whom I covered
in chapter 4. He was not a prolific published writer as many of his
contemporaries were, but there is much to be learned from him. I have
relied upon Remo Roth[33] for what follows.

In 1953, Pauli wrote to Jung, "I remarked to Bohr ... that Einstein
was regarding as an imperfection of wave mechanics (a.k.a. QP or
QM) ... what in fact was an imperfection of physics within life." Remo
Roth clarified Pauli.

> The problem of the imperfection of quantum physics
> seems to be a Schein problem (pseudo-problem or even a
> false problem). The hidden dimension behind quantum
> physics is not a physical one, but one of life itself, and to
> explore this world of wholeness we must overcome our
> one-sided causal view.

The serial multiverse takes away much of the mystery alluded
to here. As for life itself being regarded as a dimension, I think *consciousness* is the better word to use here, and it should be regarded
as being able to exist in space with a mobile energy body that can
move independently of its physical counterpart as has been attested
to numerous times throughout history and recorded in many books
on out-of-body-experiences and near-death experiences.

Pauli recognized the limitations of quantum mechanics and of
relativistic physics as well. In addition, he had a strange ability to produce a locally acting phenomenon referred to by his colleagues as the
Pauli effect. Carl Jung experienced a similar phenomenon as described
in chapter 4. It seemed that these effects were included in the same
category as another phenomenon formally referred to and defined by
Jung as a synchronicity. While this term seems appropriate for the

above effects, it appears unsuitable for Jung's experiences, for which *synchronicity* was coined. These other experiences typically involved what he termed an archetypal or iconic symbol. It occurs in a sequence with occurrences only hours apart with as far as is known up to eight manifestations, as was experienced by Jung. This was also described in chapter 4. It appears to be a-causal but not synchronous; therefore, its origin is assumed to be external with no known mechanism available yet, but it might just be traceable to the Template.

Having reached a satisfactory but far from trivial closure on Bohm's view of time, I move on to resume another thread involving time by switching back to Alfred North Whitehead with substantial help from two philosophers.

Whitehead's Description of Time and the Feist-Whitehead Interpretation

There is a complexity and hidden depth to some of Whitehead's writings on time despite his changing view of it. Hammerschmidt[34] informed readers that Whitehead had three main intervals during which he focused on certain problems. He created special words to describe his ideas, and three Whitehead glossaries belong to those three main spans of years. Although the basic list of five observable properties of time is consistent throughout these years, his interpretation of nonformal time changed, and it seemed that he did not refer to what he had written earlier. I thus find it significant and ironic that he requested that all his notes and diaries be destroyed upon his death. This is the complex background that likely contributed to there being two versions of his discourse on nonformal time. Whitehead also occasionally used the word *time* in a sort of generic sense knowing well and admitting it was not exactly what he wanted. I independently discovered this same situation.

Hammerschmidt listed these quotes from Whitehead's books.

1. "The two sides of time—the formal and the non-conceptual or non-formal—offer material for a very profitable ontological study."

2. "The presence of a non-formal side to time, or to a creative advance, has been recognized in philosophy since Plato. Yet it has, with a few notable exceptions, received little discussion, probably because it is by its nature incapable of formal analysis."

3. "The interweaving of the formal and the non-formal aspects of time is a union of a pair of opposites essential to nature."[35]

I am unclear why Hammerschmidt listed the first quote in that position; it seems to me that it is a later idea, much more profound than the others if you match it with the STZM with its two timing systems (type 1 and type 2). Moreover, what I have done in this book is exactly what Whitehead suggests would be worthwhile doing. In quote number two it seems that reference is made to the embedded time series of notable achievements made by humans—inventions and advances in knowledge recognized by the Greeks. This "side to time" is of no interest to me in this book; it was obsolete before Whitehead died in 1947. He may be justified in using the term *sides of time* for this early version of nonformal time because this involved a subjective feature in the general sequence of events that is traceable to Plato and Aristotle. In quote number three, it can be seen that Whitehead has shifted his thinking; he seems to be recognizing the existence of a second type of time that was much later termed *metaphysical* time as you will see shortly. So in what follows, the term *aspect* is dropped and replaced by *type*. But then, how do we interpret his statement that the two types of time represent "a union of a pair of opposites"? Could he have envisaged or dreamed of a mystical, Cosmic time going forward while the events embedded in his formal time continually came from the future? The differences are improved by a later quote that was not compatible with the third of the three quotes above. It represents a major shift in his thinking about time.

Quote number 4 is: "The measurement of time requires the multiple application of equal units; the geometry of time requires isomorphic regions[36]." Here we see how Whitehead first defines his formal time, which is a minimalist view, then second, his new view of nonformal time which looks as if it is equivalent to my type 1 time.

Richard Feist[37] weighed in on the updates: "In sum, and this is not a unique view, I suggest that we should read two types of time into the metaphysics of Whitehead. One type, namely that belonging to physics, is the time that is actualized by [real] entities ... let us call this physical time." I will provisionally label this type-II time, which I equate with my type-2 time in the STZM system.

Feist then defined the fundamentally different nonformal time of Whitehead, which "requires isomorphic regions." This "other sense of time, I will term metaphysical ... This would be the time that is ... connected to the eternal extensive continuum." I will label metaphysical time as type-I time, which essentially seems equivalent to my type-1 time in the STZM system. Note that Feist started by using the expression *two types of time*, which agrees with my usage. Thus, it is confusing to then use the term *other sense of time*. This should be dropped like similar subjective terms such as *aspects, sides,* and *faces* used on various occasions by Whitehead as well as other philosophers. They could conceivably still be used if purely philosophical discussions are restricted to within physical time.

What we now have is a different and much deeper type of time called metaphysical time by Feist, who drew it out of Whitehead's sketchy version of geometrical time. But where did Whitehead get this concept from? Did he get it from Dunne who in essence had a geometrical scheme involving two types of time? Regardless of origin, this time appears to be compatible with the STZM model. It would thus be appropriate to say that it is "essential to nature" (Whitehead's expression in quote 3 above).

Feist associates "metaphysical time" with the "extensive

continuum," which could be seen to have its equivalence in the torus manifold that encapsulates the universes of the STZM model, recognizing that "continuum" needs to be urgently replaced by a term that projects a finite, gridded, matrix field. Feist[38] stated that it is "our space-time cosmic epoch ... that grows ... not the ... extensive continuum." I would say that there are two main possibilities: either they both grow together, or the *extensive field* forms ahead of "our (Cosmic) space-time" string of universes. The multiverse would directly correspond to Whitehead's "isomorphic regions." I have dropped the term *eternal* originally used before *extensive* (once formed, it has no obvious need to move, so it is exempt from arguments about time). In this respect, it could just as well be considered *a priori* to exist in an absolute (Newtonian) reference frame.

Henceforth, the term *continuous* will be replaced by the term *continual* as we are by hypothesis dealing with a numerical computing environment. By using the term *bit by bit*,[39] Feist thus nicely characterized the growth of these isomorphic regions as being applicable to a computational mode. This process occurs in the Template universe, after which it is copied. I refer to Feist's version of Whitehead's "geometric" time in isomorphic regions as the Feist-Whitehead interpretation.

Feist also pointed out that Whitehead viewed the future as "merely real, not actual," and "The future lacks determinacy; it is not something that is there." These statements are obviously incompatible with the STZM model and modern thinking, which incorporates the principle of eternalism in which past, present, and future events (as far as they extend) are all real and actualized simultaneously.

Dunne and Whitehead were close contemporaries, but there is no known record of any correspondence between them, which suggests that Whitehead had issue with Dunne's 1927 book. This is understandable. But recall that Dunne was the first person to try to describe time in a geometrical framework. Whitehead's second attempt at defining what Feist termed *metaphysical time* used the specific words "the geometry of time." This would have been in the fourth decade of the twentieth century (possibly 1933) by which time Whitehead should have had ample chance to read Dunne's *An Experiment with Time*, from which he would likely have concluded that while Dunne made fatal mistakes, he was justified in the geometrical approach he took. Perhaps it was Dunne's book that took Whitehead's attention off his earlier Greek view of another aspect of physical time and to formulate another type of time altogether.

We now look at how the STZM can clarify a philosophical concept that involves time sequencing.

Reverse Causality—A Paradox or Just Another Illusion?

There is or was until it was found to be solved by the multiverse model an illusion of reverse causality (also known as a-causality) lurking throughout the philosophy of science literature. It has also been tagged as a time anomaly. It has been mistakenly associated with a precognitive dream that turns out to have been the harbinger of an actualized event in the dreamer's universe. In the typical inside-a-universe view, the dream event (referred to as the effect) occurs before the later-occurring cause, which takes place in the dreamer's universe. Here, the actualized cause is followed by the real actualized event, which is the second time the effect has been presented to the percipient. From this, it should be evident that the cause-and-effect principle (causality) holds fast in a particular universe and that one cannot mix the sequence of events (or a part thereof) seen in a dream with the sequence seen in the percipient's universe.

To see this graphically, a cutout from the cosmological model is shown in figure 5.3. It explains how the precognitive dream example cannot be used as evidence for a-causality. It would help to have understood figure 5.1 before viewing figure 5.3.

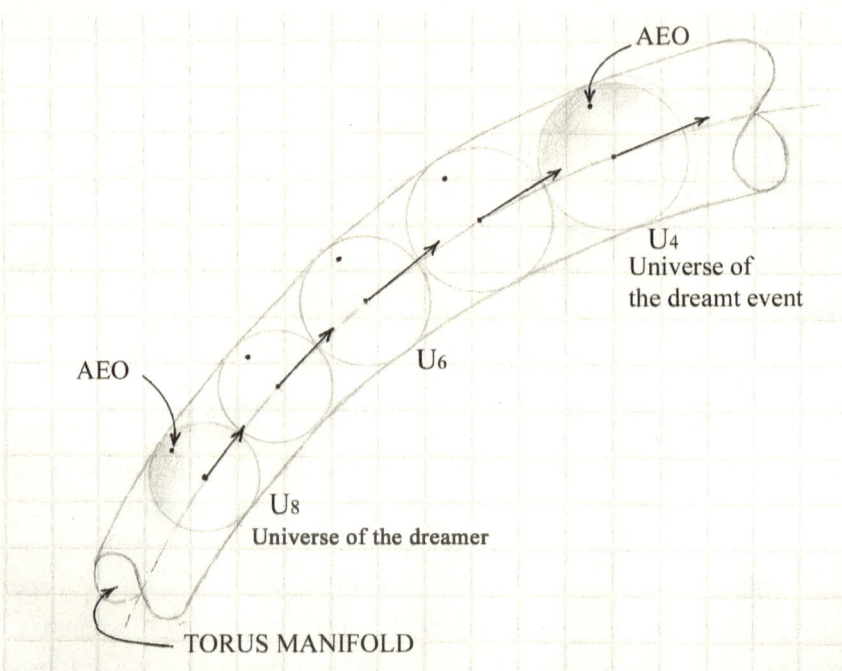

Fig 5.3 A section of the model multiverse. A group of idealized universes move, according to the arrows, inside the torus manifold. They retain their labels as they move. The dots in each universe identify the same copied galaxy containing a star with the blue planet called AEO (atmosphere-earth-ocean). The following generalized dream is based on anecdotal accounts. The dreamer, in AEO, dreams of seeing a snapshot of the stern of a large black ship perched high out of the water exposing its large rudder and three huge propellers. Many figures line the railing and some are in the act of jumping off or falling into the water. Some lights are visible on the ship. It is night time and the surreal scene otherwise seems to just hang there. Surmise that either the dreamer has connected with someone who is in this scene or someone in the scene has connected with the dreamer by way of an energetic digital 'distress signal'. The dreamer has no idea that the scene is taking place inside a universe billions of universes away in his future. I have arbitrarily labelled that universe U_4. But the scene appeared to the dreamer as if it was happening somewhere on the dreamer's planet in U_8. According to the serial time-zoned multiverse model our universe (U_8) ultimately arrives in the exact position that U_4 occupied in this diagram but to catch the whole scene many other positions either side of it have to be traversed. It is then that the Titanic saga, which occurred on the night of the 4th and 5th April 1912, is actualized in this universe two months after the dream. The dreamer is subsequently informed about the details of this event from newspaper accounts and thus becomes cognizant of the *cause*, and a second, much fuller, account of the *effect*. In this view of *precognition* it is clear that *we move towards our future*, just as Guyau claimed.

Two Complementary Views on How We Think
We Relate to the impending Future

In the following, I take the example of the different views that existed between Guyau[40] and Whitehead (see chapter 4). Do we really move toward our future, or does our future move toward us?

In Guyau's *The Origin of the Idea of Time*, written prior to 1888, we read, "The world is seen to be a world of events, coming into being and passing away … We misconceive the future if we think of it as that which comes towards us; it is rather that towards which we go."

In Whitehead's book *Adventures of Ideas*, his seemingly opposite view is expressed: "Cut away the future, and the present collapses, emptied of its proper content. Immediate existence requires the insertion of the future in the crannies of the present." This implies that the current Now moment is being updated only where essential changes have occurred since the last Now. Because everything in the entire universe is continually changing right down to the decay of radioactive isotopes at the nanosecond level, this view is not supported. Thus, the complete incoming Now from the future simultaneously (or nearly so as in a real computer) replaces the entire outgoing Now that is immediately copied to the universe behind (active past). This is the principle of the COPI procedure used by the Cosmic computer (see figure 5.1). So the second part of Whitehead's statement has to be replaced by something like this: "Continual existence requires the immediate insertion of the future Now into the space just vacated by the old Now."

If Whitehead's modified view is accepted, we can see that Guyau's and Whitehead's combined views are seen to be equivalent to the observation already made in figure 2.1B, where at any location on the multiverse circuit (e.g., labels 1, 2, 3) the same event (or slice of an event) always occurs. This is because the forward march of the universes is in essence countered by the backward-moving paste-in of the new Now from the future by the COPI procedure.

Up until now, it has largely been a philosophical issue even bordering on a psychological one. Neither proponent derived his deductions in a formal way, nor were they able to interact and create a dialogue on the issue. Thus, it is quite unclear as to how they arrived at their views. Secondary sources do not mention this aspect. In analyzing the problem, I resorted to the STZM Cosmological model ignoring the change in size of the universes with distance and duration, which are irrelevant variables in this application. The problem can be clarified considerably by making use of a time-lapse diagram simulating a sequence of moving, tagged serial universes.

Figure 5.4 illustrates Guyau's view, bearing in mind that it deliberately accentuates one side of the picture.

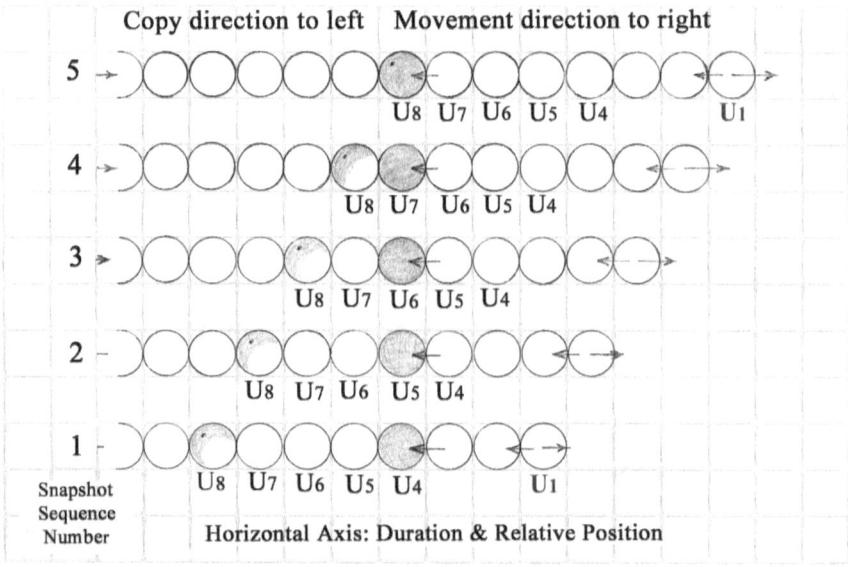

Fig 5.4 Numbers running up the left side of the diagram identify *serial snapshots* of the universe array led by U_1 which is continuously spawning new events that get copied *to the left* —counter to the movement of universes going in *micro-steps to the right*. In this sequence, subsequent rows of universes seen *above* 1 are *cut and pasted over each update*: thus row 2 replaces row 1, 3 replaces 2 etc. (There is no actual movement upwards in this diagram!). An enormous sequence of Δt's occurs between each number (1 to 5). The actual depiction of universes should look like what is seen in **Figure 5.3** but for ease of drafting I adopted the straight configuration of the *Dynamic Events Line* in **Figure 2.1A**. The *directional sense* matches those of **Figures 2.1B & 5.3**. Now's exist in the above universes, which was the case in **Figure 5.3** and all other such diagrams. In the following exercise, you have to be aware of *the whole array at once*. Start in snapshot sequence 1: the dreamer in U_8 dreams of a snapshot of the sinking Titanic in U_4. Now observe the

relativity of *motion* that influenced Guyau and Whitehead to make seemingly different statements about the future. Begin sliding your *left eye* up along the U_8 *diagonal* and your right eye up the column U_4–U_5–U_6–U_7–U_8 and thereby visualize the convergence of U_8 with the grid square which once held U_4 wherein the Titanic snapshot dream occurs. This provides the view of Guyau in which we *move toward* the future. Whitehead's view is encapsulated in **Figure 5.3** where the dynamics of the Now constrains events to travel *counter* to the direction of motion of the universes. The COPI arrows originating in the Template (and pointing left in the above figure) seem to show (relatively) that our future is moving *towards* us.

In figure 5.4 is an enormous number of incremental movements (every Δt) involved in any one universe moving from a specific position (say a grid square) to the position occupied by the universe in front of it (i.e., the next grid square). The diagram as marked up focusses Guyau's view particularly well. Whitehead's view of the approaching future is best seen in figure 5.1, which shows how the future moves toward us. This shows graphically how any event remains fixed in position on the circular trajectory line shown in figure 2.1B.

A Holistic Perspective on the Purpose of Life

Synopsis

This is another unexpected spin-off from the discovery of the serial multiverse, but as usual, there is nothing new under the sun. Had it not been for a very private questioning about my existence when at a single-digit age ("Why am I me?" and later "Why am I here?" [41]), had it not been for a book in our home library titled *The Philosophy of Life* by Anderson M. Baten (a gift from my future mother to my father in 1937), and had it not been for the fact that I have just read some familiar words written by astronaut Edgar Mitchell[42], I might not have been equipped to deal with the subject in this section. I had by then realized that I was far from being alone in my bewilderment about our existence.

Mitchell wrote that by 1967, he had become dissatisfied with the inability of philosophy and theology to satisfy his questions "about the meaning of life and man's place in the universe[43]." Of course, we are specifically focusing on human life in this section and interpreting it according to the Blueprint, which nevertheless is by its nature idealized. The concept of copied universes and the issues of free will and determinism give a broad clue to answering Mitchell's first question: just what is the meaning of life? The even broader question of humanity's place in the universe seemingly overlaps the first question but not entirely. It is now necessary to look at another facet of free will. These facets are the reason why the topic of free will keeps recurring in this book.

Some Other Perspectives on Free Will

The issue of whether we are governed by determinism or are responsible for our actions by making freewill choices has ramifications in a number of areas so far not mentioned. The particular structure of the multiverse derived in chapter 2 and the inaugural but limited interpretation given in chapter 3 project a strong ontological link. Therefore, it is worth expanding on free will in this light. I start by quoting the beliefs of psychologist Susan Blackmore, who bluntly stated, "We do not have freewill … Many people argue that life, law and society would be impossible without freewill, or at least without an illusion of freewill, but I disagree."[44]

Consider a lawyer's view on the issue.

> Freewill and determinism are said to constitute the most written-about problem in the history of philosophy, one that continues every year to produce major publications by major university presses, each claiming insight into moral and criminal responsibility … I argue, in contrast, that the relationship between freewill and determinism is a false problem: that is, a problem that we are incapable of resolving, even in theory … The proper response to a false problem is not to search for further evidence or to strive for better analysis. The proper response to a false problem, including that of freewill, is to stop thinking about it.[45]

Interpretation of the STZM model puts paid to that argument.

My deduction is that in this universe, we mostly do not have free will, and where it might be claimed to occur, it is a local, soft, free will experienced by only certain people. Apart from the Template, the cases that we have seen—an example being the *Titanic* effect—the fundamental copy procedure of the multiverse tends to overwhelm it. In addition, our societal structure tends to block it out for most people even in the Template.

But don't panic, because of this and what Blackmore has said. As I determined from a poll on this issue, the very common belief that we have some free will (and the rest is predetermined) should get us through life. It is possible to just go along with the illusion of free will and still be happy. It is ironically the same advice that lawyer P. K. Westen recommended but based on different reasoning. In leaving it there, I will avoid discussing the obvious legal ramifications.

I think that if people retain some degree of clinical mindedness, accepting determinism is little different from watching what actors do and say for example in a Shakespeare play. For those who are not so clinically minded, this advice might just be unwanted bad news. The big-picture antidote for this is to be aware of the actual nature of our existence: our souls are the reason why we are living this particular life in the first place. I agree with many things that Bernard Haisch has written, but in his second book[46]), he developed a strong rejection of determinism.

> If we think of ourselves as just complex organic machines, then every thought and every action that we think we are consciously and freely choosing is in reality just the consequence of the previous state of the universe … There is no freedom … no choice.

Later, he rephrased the same idea: "In this view we were born with every thought we would ever have in our lives pre-programmed … like a CD playing from start to finish … This is a truly depressing view of our nature."

Here we can see here how the human view of existence is distorted if we adhere to the belief that we live in a single universe. What takes place in the multiverse Template now would be all there is. In principle it can exist in its own right: its descriptor would be just monoverse. But the evidence just doesn't support this view, and as I have pointed out, even there, free will for most persons is very hard to get. This is because of the pyramidal control structure that develops naturally in human societies.

In such a monoverse, no one could see the future. That violates what is actually known to occur in this universe as discussed extensively throughout this book. Just why are there now eleven multiverse proposals? This level of activity signals a largely twenty-first-century revolution.

I am not totally convinced we are really supposed to know our free will has been compromised. While many lives are pleasurable for the soul, many others are "truly depressing"[47] with every stage in between. Most people tend to meet the challenges of a life head-on, day by day based largely on what happened in the Template. Now ponder this: it was possible for me to get to this stage of thinking simply by using the STZM. Considering that my motive to build it was just to explain a single phenomenon—precognition—the STZM has taken me a long way further.

Haisch had enthusiastically discovered the perennial principle, which is no news to spiritually tuned people the world over though it is not necessarily known by that name. The choice of lives (provided by physical bodies) in the STZM model is clearly immense. F. A. Wolf[48] is another physicist who is comfortable talking about our basic spiritual existence. There are two other books with the title *The Conscious Universe*[49] that seem to be complementary to the above, but they actually take quite different approaches to the concept. You will have noticed by now that this situation is almost becoming a theme for this book.

These recent explorations of various aspects of consciousness were long preceded by a classic book by Frederick Myers [50], who communicated with spirits of departed souls to establish that consciousness survives bodily death. This was clearly seen later when messages began to be received from Myers's spirit after his bodily death. It was possible to confirm it was Myers' spirit by the nature of those messages which were split into parts and sent to three different widely separated psychic recipients and then recombined by friends who knew Myers intimately.

Actors in Deterministic Roles: No Metaphor

The elements of the illusion of free will may be seen by studying those who act in plays. We have to back up four centuries to find the right lines. In 1599, Shakespeare introduced a well-known and often quoted monologue in *As You Like It*. The beginning three verses tell us, "All the world's a stage, And all the men and women merely players; They have their exits and their entrances."[51]

I truncate it there because other deep messages likely appear in the full monologue and I am fearful of diversions. Is this just a metaphorical comparison between real life and the world of acting on the stage, or is the message much more profound?

I searched the Internet for people's interpretation of the above lines.[52] Following is a précis of two of the very different answers: (1)"It's basically just a metaphor about life" and (2)"It means we all have roles to play. We like to think [that] we are independent and that we choose how we act, but in reality we are acting according to scripted roles."[53] This last statement is clearly in support of a deterministic human existence, which is the appropriate interpretation for everyone in this universe.

While on the topic of Shakespeare, I would like to suggest an update of a verse in *Romeo and Juliet*: "It is not in the stars to hold our destiny but in our-selves" can be replaced with a more fundamental statement: "It is not in the stars to hold our destiny, but in the universes ahead."

The next subsection is intended to show that more individuals than Wolfgang Pauli can accept natural scientific phenomena and paranormal (or preternatural) phenomena together. The following individuals all used careful observation and recording techniques in both fields as if they warranted equal attention: Charles Richet, Camille Flammarion, Alfred Wallace, Arthur Eddington, and Jung. Pauli has already been covered in chapter 4, but he was not an experimentalist. He was a theoretician. Regrettably, Dunne has to be excluded because of his refusal to reveal his spiritual life until he wrote the manuscript for his last book, which was published posthumously. Following are examples of some of the individuals who seemed to naturally accept the two apparently disparate areas of research.

A Shadow Character: A. R. Wallace

In the nineteenth and early twentieth century there lived a remarkable man who was looked upon as being in the shadow of Darwin, but in some aspects, it was he who cast shadows on Darwin as well as many other contemporaries. He was Alfred Wallace[54] (1823–1913). Much praise was bestowed on him when he was working in the area of organic evolution, but he had another side to his interests. He was also involved in research in the preternatural areas. In doing this and in the company of Charles Richet, MD, and other notable Frenchmen of his generation, they were anticipating Pauli, Bohm, and others in trying to fulfil the pervading belief that a complete theory of human life on this planet must embrace not only the physics (including scum chemistry) of the whole universe but also our interactions with phenomena that clearly lie outside established science.

When it came to his protocols, Wallace rigorously studied the natural and the preternatural phenomena as if they were all worth equal attention. The Victorian era was not a particularly good time to do that. He did not start his own preternatural investigations until after becoming convinced by family members that it should be taken seriously. Eventually, he believed that the raison d'être of the universe "was the development of the human spirit" though he may have come to this specific view in his association with the spiritualist communities he had infiltrated. An equivalent statement on the matter came later from Teilhard.[55]

Edgar Mitchell's Other Question

I will try to answer Edgar Mitchell's second question in this short subsection though it is worthy of a section of its own. He asked: what was "man's place in the universe." My angle on this, expressed earlier, is that we humans were deliberately formulated to exist at a scale that positions us about midway between the very small subatomic scale and the very large cosmological scale.[56] This positioning was done, I conjecture, so extremely advanced physicists, engineers, and quantum computer individuals (who are on the rise) did not cause havoc with the machinery of our universe—for their own good. The best-known example, so far, of causing local havoc to the machinery at the micro-small scale is the control over the nuclear fission process, which has reached the meddling hands of far too many nations already.

Because I have identified more than one variety of free will, it is now necessary to return to that perennial topic once again.

The Free Will Issue Revisited: Another Facet

The topic of free will keeps reemerging from the background of developing concepts, especially those occurring in this chapter. I have avoided trying to control this because so many of these intrusions are caused primarily by my swinging the spotlight on different people. Each person exposes a different facet of the topic.

As you saw earlier, Bohm, Blackmore, and now I (with some caveats) believe we do not have free will, but our individual reasons for this conviction are quite different. My reason is based on having demonstrated that we live in a copied universe in a geometrically based Cosmological model. Bohm's argument comes from a background of physics (especially quantum theory), logic, and philosophy.

In contrast, Blackmore's reason seems to reside mainly in intuition and psychology.

Bohm's reason is not altogether straightforward and makes for an interesting discussion. He wrote, "If you live your life considering that you have freewill, then the contents of the world will appear fragmented and life will appear to be driven by freewill[57]." For humans living in the Template when it was at our present stage of development, their existence did contain free will—the amount depending on where they belonged in the societal structure, which is an entirely different situation. Because their inorganic universe was seen to be deterministic, being based on the established laws of physics, their view would have been that they had episodic free will mixed with various amounts of determinism from different sources. They would have been correct in saying that, and if these Template people are still in existence, their statement should still apply.

In our copied universe, Bohm is saying in effect that according to his view of the universe, everything—organic and inorganic—is connected and therefore there is no room for free will. However, there seems to be quite a stretch made in arriving at such a statement. Nevertheless, it does agree with my conclusion reached via another pathway. To avoid misunderstanding, here is Bohm's flip side statement: "If you live your life considering that you do not have freewill, then the contents of the universe will appear coherent and life will (appear to) be driven by deterministic forces[58]."

If you still insist that you have free will, you have been successfully caught up in the illusion of free will, which I deduce is exactly how it is meant to be because if we are primarily spirits seeking a human experience, it is logical that our souls should be treated to lives that give them exposure to the lives as nearly as possible to the corresponding one in the Template.

In the next section, I present some anecdotal accounts that originated from well-known people who lived in the early part of the last century (see appendix 2). The experiences strongly indicated that there was something seriously lacking in the then-current understanding of time and that this was apart from the situation caused by Einstein's theory of relativity. Most of the evidence was then coming from dreams. Einstein's physics did not deal with dreams of the future despite the fact that precognition is a manifestation of the relativity of events.

Some Notable Accounts Concerning Sequence Order Anomalies

Determinism Lurking in the Pages of the *Times Literary Supplement*

Herein you will be introduced to new characters some of whom interacted with characters already featured in chapter 4. This will show you that the web I was spinning is much more extensive than the one I presented earlier. In a sense, it illustrates a facet of Bohm's holistic approach to our existence.

An astute review by J. W. Sullivan[59] of Dunne's first book appeared in the *Times Literary Supplement* (*TLS*) of London in late 1927. After praising Dunne for his account of the dreams, an opinion which I endorse, Sullivan made it clear he was quite skeptical of Dunne's geometrical model of time. He summed it up: "But, whatever we may think of Mr. Dunne's own philosophy of time it is certain that something almost equally strange is necessary to account for his results[60]." I wonder if there will be a politely skeptical reviewer who writes something similar about *Time and the Multiverse*.

Seven days after Sullivan's review, further correspondence appeared in the *TLS*. The review had prompted the famous English Egyptian archaeologist Flinders Petrie[61] to openly narrate two anecdotes.

> My [late] cousin Dr. John Bromby ... was living with his sister in England. She described to me how, one morning at breakfast, he told her of a ridiculous dream about a mouse coming out of an unexpected place, running

about the room, and being ignored by the cat in a strange way. Before they [had] left the table the mouse suddenly appeared; they watched it carefully and the whole episode went forward exactly as he had described it. Both persons were not at all superstitious, rather Sadducees than otherwise.

Petrie then recounted a story of prescience he had read in a book written by William Dampier published in 1729.

The Darien natives ... predicted to ... a ship's doctor, Lionel "Waper" (I later discovered that it was a misprint of Wafer) ...that ten days hence, two ships would arrive, one being English; that one of his party would die and one gun be lost in going on board ... All of this fell out exactly according to the prediction[62].

Petrie then stated that this story "ought to be taken seriously." He cryptically summed up the situation regarding time and the ability to see the future with the caution, "We must begin to learn the world over again before we know where we are." I suspect that this may have triggered Dunne to write words that echoed this view: "If prevision is a fact, it is a fact which destroys absolutely the entire basis of all our past opinions of the universe."[63]

I miraculously located a photocopy of Wafer's 1696 book printed in old English script. It added more impact than the highly abbreviated fourth-hand account via Petrie. The most striking aspect is that the native shaman and his assistants collectively called *pawawers* made an astoundingly detailed prediction in answer to a single very straightforward question from the buccaneer group, of which Wafer, the ship's surgeon, was a member. The buccaneers would have thought that the natives probably had a rough memory of the frequency of arrival of ships in the area and simply wanted an estimate of when the next ship might arrive. They got from the obliging pawawers albeit not without a struggle much more than they had bargained for! This seemed to be a new experience for Wafer. It certainly impressed the better-known Dampier enough for him to reprint the full story in his book.

It shows that directed prescience (in a non-dream but probably trance mode) perhaps equivalent to modern-day remote viewing (a term due to physicist Russell Targ) can be assumed to be widespread geographically and historically. Notice that I deftly avoided the use of the word *time* to show that in most cases, it is possible to do so. Events are the key to doing this as Whitehead importantly pointed out.

The *TLS* literary saga now resumes with a familiar character reappearing. On Thursday, October 13, 1927, R. L. Mégroz[64] wrote to the *TLS* pointing out quite strongly that "The anticipatory dreams he [Dunne] describes are not at all new as an experience" and not new concerning "detailed records" either. He claimed to have a mounting collection of such stories that covered the "last few centuries," which included stories from friends and notable people "of our society." You may recall from chapter 4 that he wrote to T. S. Eliot about obtaining just one of TSE's dreams. In that quest, he was disappointed. He also made a request for readers of the *TLS* who have had "remarkable" dreams with time anomalies as the main element to communicate with him. He was planning an anthology of such dreams. He further wrote disapprovingly,

> Early last century [Heinrich] Heine [the journalist and literary critic] declared that precognitive dreams were a peculiarity of modern civilization and that the great ado about dreams made by the ancients, both civilized and savage, was a sign of the comparatively unusual nature of vivid dreaming.

Appendix 4, based largely on anecdotal accounts listed by J. B. Priestley in his book *Man and Time*, shows just how serious a business it can be gathering data on apparent time anomalies experienced by many people, in this case over 1,000 in the British Isles. The majority obligingly and sincerely wrote to Priestley about their experiences or those of others known to them. Priestley's management of the data (necessitated by the diverse types of anecdotes in these letters) is very instructive and possibly original in its field. There were others who independently collected such data.

Camille Flammarion,[65] a French astronomer and psychic researcher, had collected over 4,500 letters containing anecdotes of what would be considered paranormal events; many contained accounts of precognition. Letters came from as far afield as Argentina. To top all this off, Robert Nelson, a circulation executive with the *New York Times* established in the early 1970s the Central Premonition Registry with the collaboration of Dr. Stanley Krippner, a psychologist and parapsychologist then at the Brooklyn Medical Center. In receipt of over 5,000 letters, they discovered that just below 1 percent of the anecdotes qualified as successful hits. Krippner reported that humans "dream very morbid terrible things" such as injury and death to prominent people, accidents in general, fires, floods, and plagues. I would conjecture that this must surely be due to the energies of those events (and in people surrounding the events) that must cause something analogous to signals in carrier waves in the energetic field. By *analogy* with AM or FM radio waves, these energetic waves must contain encoded information.

I end this section with a mention of the theologian, metaphysician, and consciousness researcher James van Pelt[66], who expressed deep concern with the current situation regarding the practice of suppressing accounts of persistent anomalies "because they cannot be accounted for by the prevailing model of reality." This is much like the lament of one of Priestley's pre-1964 correspondents who decried the general avoidance of funding research areas that would enlighten us on matters of our very existence (the field of ontology). Here is how van Pelt expressed his frustration.

> The prevailing model of reality is equipped with anti-explanatory power: a filter to neutralize the threat represented by any rival belief system by waving away any empirical or anecdotal evidence that could violate it,

including any sign or revelation erupting into everyday experience[67].

The reality is that the "prevailing model" isn't equipped to deal with preternatural phenomena at all; the "waving away" of paranormal data and analyses is done in a different venue by the standard model users or their commentators. This is where the champions of the past holistic approach to our existence (e.g., Whitehead, Pauli, and Bohm for example) should be lined up and requoted as well as updated by others of the same mind by representing a block defense of preternatural phenomena, that it will not go away, despite the opposition's taunting. After all, there are indications that quantum physics contains weird characteristics that cannot rationally be explained.

More Reflections and the Deepest Mystery

On the last page of his book, Camille Flammarion wrote the then startling words "Yes, the unknown World is vaster and more important than the Known."[68] He was of course referring to the spirit domain. In stark contrast to this, our material world is overloaded with mysteries of its own—in quantum physics, which has been modified or reinterpreted several times, and in relativity theory, which has led to the nebulous hypotheses of dark matter and dark energy and black holes. These theoretical areas are in turn strongly coupled to burgeoning experimental disciplines such as: astronomy, astrophysics, and cosmology, which become full of new discoveries that spawn hypotheses—even metaphysics.

These particular areas of scientific research are where venturous scientists are to be found in very great numbers; they are at the frontier of big science. No wonder there is a dearth of manpower in the fields of the preternatural! Sir Roger Penrose noted, "One of the deepest mysteries of our universe is the puzzle of whence it came"[69]. The cosmological model developed in chapter 2 is set up to push this mystery a step back and allow us to say, "The deepest mystery of our multiverse is the puzzle of whence it came." Penrose made some pertinent statements that seem to me to reflect aspects of what David Bohm[70] had written three decades earlier. Note that the following material must be a metaphor for the real thing.

After pointing out that a very precise clock (so to speak) is needed to implement the equations of physics, he concluded that a cosmic clock was required, and in this context, the whole of the contents of the universe must be linked. First, these statements need to be recast in the context of numerical computing rather than analytical equations, and second, not surprising, here is an idea for a clock in a cosmic computer or more exactly, as we have seen, a Cosmic quantum computer with a timing device. Does this seem to support Bohm's conceptual hypothesis?

My attention was also drawn to an essay on time by Julian Barbour[71] whose book[72] I found extreme in its determination to subjugate the role of time. Despite having difficulty with the basic approach to time in that essay, I will capitalize on a particularly graphic concept that appears in it. He asked: "how billions upon billions of natural clocks scattered through the vast reaches of space can tick in step[73]." Does this now seem to be another perhaps metaphorical reference to cosmic quantum computer clocks inside other universes? At this stage of speculation, I will have to hold on to the Cosmic computer concept with its function to drive the multiverse, overall.

One thing is quite clear: I have been restricted to conducting my thinking in a system of metaphysical mechanics based on plain Euclidian geometry *sans équations*.[74] Regarding this last aspect, I find myself in the same camp as many qualified cosmologists from different disciplines who are forced to work under the same or similar restrictions. However, it turns out that there are two key areas that usually prevent others from advancing very far into constructing a large scale Cosmology as shown by figure 2.1B. These areas are seeking a proof of the existence of a multiverse model, which can evidently be done in at least three different ways, and taking seriously the belief that our immortal spirits (temporarily existing as souls) are the mainstay of a dualism, the other part of which is the transient physical bodies that are easier to dispose of than throwaway cameras containing mercury batteries.

Regarding the first area, the three ways of demonstrating that a multiverse exists are

a. starting from the block time concept, transfer to static block Nows as representing finite slices of events and then animating the system and search for data to verify it,

b. start an argument based on precognition and retrocognition that acts as support for (a), and

c. conduct an experiment (which has already been considered by others but which I do not know is practical) using a universal quantum computer that communicates to other universes and seek a report back that something is out there.

If that something acts like our universe and it is connected to another universe like ours, etc., that implies that we exist in a many-worlds or multiple-universe group. Such a group—the MIW concept—has been postulated to exist based on a current reinterpretation of quantum mechanics (see chapter 6, essay number 4.)

I end this section by referring to the most recent but not always so serious book in a series edited by John Brockman.[75] Here is an opportunity for big-name scientists to have the temporary luxury of airing their beliefs without proof and without references. In it are numerous assaults on established ideas in physics, including a long-standing relativity favorite—space-time.

One skeptic of multiverses is Lee Smolin, who stated,

> Explanations such as the anthropic multiverse ... are unscientific because they call on unobservable collections of other universes and make no predictions by which their hypotheses might be tested and falsified.

> There is … a chance for science to answer these ques-
> tions, and that's if the Big Bang was not the first moment
> of time[76].

Disallowing big bounces, the STZM model has the right features to address Smolin's conjecture about the first "moment of time". Moreover, it is readily testable as I have shown in the majority of the pages in this book. Indeed, the STZM can be appropriately described as an anthropic Cosmology.

Notes

1   http://en.wikipedia.org/wiki/Cosmic_time.
2   D. Bohm, *The undivided universe: An ontological interpretation of quantum theory*, B. Hiley, trans. (UK: Routledge, 1993). I used a book review to get the essence of it: J. Leslie, *The Absolute Now* (in *London Review of Books* 16(9):15–16, 1994). According to Leslie (a philosopher), Bohm and Hiley, both at Birkbeck College, worked on the theory of the undivided universe.
3   D. Bohm, *Wholeness and the implicate order* (London, UK: Ark Paperbacks, 1983). It was first published in 1980 by Routledge and Kegan Paul. The amazing thing is that you get his explicit and revealing idea about time on just two pages—211–212—of a 213-page book.
4   D. Bohm, *The undivided universe: An ontological interpretation of quantum theory*, B. Hiley, trans. (UK: Routledge, 1993). I used a book review to get the essence of it: J. Leslie, *The Absolute Now* (in *London Review of Books* 16(9):15–16, 1994). According to Leslie (a philosopher), Bohm and Hiley, both at Birkbeck College, worked on the theory of the undivided universe.
5   D. Bohm, *Wholeness and the implicate order* (London, UK: Ark Paperbacks, 1983). It was first published in 1980 by Routledge and Kegan Paul. The amazing thing is that you get his explicit and revealing idea about time on just two pages—211–212—of a 213-page book.
6   P. Davies, *About Time* (New York: Simon & Schuster, 1995). A short version of the Now problem is discussed on page 77.
7   J. Barbour, *The end of time* (Oxford, UK: Oxford University Press, 2000). He said, "The block universe picture is in fact close" to his theory. Thus, it is an incomplete treatment without some form of built-in but artificial motion implied. This is supplied in the serial multiverse model, which is associated with Newtonian mechanics. The Now concept is explainable only in

a Newtonian setting rather than being entangled in a space-time setting. I have quoted mainly from page 142 of Barbour's book. He has evidently not stressed Whitehead's ideas on events as being more fundamental than time.

8   R. Carnap, *Carnap's intellectual biography*, in P. A. Schilpp, ed., *The Philosophy of Rudolf Carnap* (1963), 3–84. La Salle, IL: Open Court—extracted from *the Stanford Encyclopedia of Philosophy* article "Being and Becoming in Modern Physics" by Steven Savitt, copyright 2013.

9   Ibid.

10   MIW has been variously described as a model, an interpretation, a view, and an approach (☺).

11   See en.wikipedia.org/wiki/Multiplexing.

12   Boston University's Tommaso Toffoli is a professor of electrical and computer engineering who has been on the faculty since 1995. He is known for the (intelligent) design of the Toffoli gate, which can be incorporated into the design of a quantum computer. At MIT, Professor Seth Lloyd is also working in this area.

13   Taken from W. W. Hammerschmidt, *Whitehead's philosophy of time* (New York: Russell & Russell, 1947). Alfred North Whitehead ran up against the same problem I experienced. Hammerschmidt makes this clear by saying that the use of the single word *time* "is purely an arbitrary usage which [ANW] imposes to facilitate discussion of the subject." With the multiverse model in mind, I can begin to see that Whitehead and later Bohm had grasped the essential elements needed to develop a theory of our existence. Both men avoided getting embroiled in the psychological issues surrounding time. However, the quote from Whitehead (1933), in which he says that the future moves toward us is among the most profound things that he wrote and perhaps anyone wrote up to that time in regard to understanding the Now problem. Though the same holds true for Bohm (see his footnote 3), unlike the latter, Whitehead relied on process philosophy whereas Bohm used fundamental processes expanded to deal with the whole universe. Insofar as material about Whitehead is available on other sites, my main source has been the *Stanford Encyclopaedia of Philosophy*; http://plato.stanford.edu/entries/whitehead/ (revised October 1, 2013). [Contribution by Andrew David Irvine: andrew.irvine@ubc.ca]

14   The section title is taken from Giuseppe Verdi's opera of the same name. Verdi was no stranger to this force.

15   http://www.worlddreambank.org/L/LINCOLN.HTM.

16   I took most of the information I used in this story from a site maintained by anomalyinfo.com because it was accompanied by source material. They based their clip on the book by A. B. Paine, *Mark Twain: A Biography*,

volume 1 (New York and London: Harper & Brothers, 1912), 133–43. The anomalyinfo.*com* reporter had posted a note that Paine was the first one to document the dream and the actualized death scene together; nothing about them had been published before, but the sources were quoted as being in diaries and other paper documents. He expressed the fact that we can rely only on Paine. This story isn't falsifiable, and there are a number of recent authors who still quote it. Perhaps significantly, in 1884, Clemens joined the American Society of Psychical Research when it was established. This was an offshoot of the British SPR created two years earlier. Clemens's interest in the ASPR is on record, but it covers only his observation that authors of books frequently seemed to converge on a topic or idea at about the same time, and it was his belief that it was by telepathy. There is no report on record of his 1858 dream, and this is to be expected. Apart from the fact that the dream was by then twenty-six years old, the SPR rules for registering precognitive dreams are so stringent that the SPR deemed even Dunne's remarkable precognitive dream anecdotes were improperly documented, and yet to my knowledge no one has objected publicly to his dream reporting. My own precognitive dream record wouldn't pass the SPR either as I did not have a witness.

17   M. Robertson, (1898), *The wreck of the Titan* (A novella in the book *Futility* published by M. F. Mansfield). It was reproduced in the book referenced in the table below. The *Titan* had close similarities in specifications to the *Titanic* as well as to several events that befell that ship. The following table gives a comparison list of close correspondences taken from various sources. See also http://en.wikipedia.org/wiki/Futility,_or_the_Wreck_of_the_Titan.

A number of features of Robertson's *Titan* did not conform to the *Titanic* such as it having sails on two masts. If Robertson was dreaming about the *Titanic*, the results suggest a mixing of past images with predominantly future dream images; this is a feature Dunne made a point of explaining as a result of analyzing his own dreams. I assume that Robertson had a series of dreams about the *Titanic*; it is well known that dreams can repeat, an example being Heinz Pagels's recurring (truly predictive) mountain climbing dream covered later in this chapter. No one has been successful in proving that the statistics for the *Titan* were faked.

There is one other observation not included in the table. I suspect Robertson might have initially named his vessel *Titanic*, but to avoid negative or even legal consequences with the White Star Line, he dropped the *ic*. That shipping company had put into service a vessel called the *Majestic* in 1890. It did not have sails, but steam-driven vessels with sails still existed in 1898, so there is the possibility that this was another smoke screen to deflect attention from the White Star Lines. The sails on the *Titan* are deliberately not mentioned in the following table as not being critical in the comparison.

Table of Comparisons between the Statistics of the *Titan* and the *Titanic*

Titan 1898	*Titanic* 1912
1. Triple screw	1. Triple screw
2. Practically Unsinkable (or equivalent), due to watertight compartments.	2. "Designed to be unsinkable." Later called "practically unsinkable."
3. Length—800 feet	3. Length—882.5 feet.
4. Displacement—75,000 tons[+] Wiki gives gross weight as 45,000 tons[*]	4. Displacement—63,000 long tons. Lord[++] gives 46,328 gross tons.
5. Largest craft afloat.	5. "World's largest luxury liner."
6. Twenty-four lifeboats; less than half that required to save a full complement of 3,000 people.	6. Sixteen lifeboats and four folding boats, less than half of those needed to save a full complement of 3,000 people.
7. Traveling at 25 knots at time of impact.	7. Traveling at 22½ knots at time of impact.#
8. Position~400 nautical miles from Newfoundland.	8. Position~400 nautical miles from Newfoundland.
9. Hit iceberg on starboard side.	9. Hit iceberg on starboard side.
10. Disaster occurred on an April night.	10. Impact with iceberg on April 14, 2012 at ~11:00 p.m. and sank out of sight at 02:20 on the 15th.
11. The Atlantic crossing was west to east.	11. The Atlantic crossing was east to west.
12. More than half of her 2,500 passengers and crew perished.	12. More than half of her ~2200 passengers and crew perished.
[+] Ian Stevenson, MD, "Paranormal experiences connected with the sinking of the *Titanic*," in *The doomed unsinkable ship: The wreck of the Titan*, W. H. Tantum, ed. Riverside, CT: C'S Press, 1974).	[++]Lord, W. (1955). *A night to remember*. (Mattituck, NY: Amereon House. Reprinted 1987 by Holt, Rinehart & Winston.)
[*]Taking a modern empirical graph of length versus displacement for large steel tankers, an adjusted weight for second item 4 if the length was 882 feet would be about 68,000 tons. This is far too high as I expected. The tonnages given in various articles do not always agree.	#The stated maximum speed was 23–24 knots. Contemporary Cunard liners could go a knot or two faster than this. But Stevenson[+] said the *Titanic* could do 24–25 knots; elsewhere 23.

A concept that I have called the *Titanic* effect (TE) takes its name from the *Titanic* drama. Though the records show there is overwhelming support for ignoring precognitive warnings, there may be a few cases in which people have claimed that heeding the warnings saved their lives. This is basically the TE. Note also that when discussing soft free will, I hypothesis that a hypothesized critical personal energy level may be exceeded in rare cases resulting in an override of personal determinism; it would be called a local effect.

In his novella, Robertson had people and a drama on the iceberg itself. On the night of the *Titanic* sinking, officers of the North German Lloyd liner *Princess Irene* reported sighting an iceberg at the foot of which were the bodies of more than a dozen men wearing life buoys. They were huddled in groups and fifty miles from the scene of the sinking. http://m.thechronicleherald.ca/titanic/archive/82908.

18   Ibid.

19   *New York Times*, April 23, 1912; see also https://www.encyclopedia-titanica.org/stead.

20   http://www.boundaryinstitute.org/bi/premon911.htm.

21   http://www.visionaryliving.com/. I can answer that question quite easily by saying that it is a clue to the structure of a multiverse in which our universe is embedded. Obviously, many souls take that clue with them to the great beyond on the other side.

22   L. Dossey, *The power of premonitions* (New York: Dutton, 2009). This author includes the term *precognitions* under *premonitions*.

23   http://www.visionaryliving.com/. I can answer that question quite easily by saying that it is a clue to the structure of a multiverse in which our universe is embedded. Obviously, many souls take that clue with them to the great beyond on the other side.

24   L. Dossey, *The power of premonitions* (New York: Dutton, 2009). This author includes the term *precognitions* under *premonitions*.

25   H. Pagels, *The Cosmic Code: Quantum Physics as the Language of Nature* (New York: Simon & Schuster, 1982). In it, he mentions his idea "that the universe is a giant computer" and the fact that his dream was repeating. Discussion of them is given at various locations in the text. I quote also from the following: http://www.nytimes.com/1988/07/26/obituaries/dr-heinz-pagels-49-a-physicist-dies-in-fall-from-colorado-peak.html.

26   Ibid.

27   Ibid.

28   http://www.supertopo.com/tr/Pyramid-Peak-Paul-Ryan-s-Favorite-Fourteener-or-Fitness-Fable/t12106n.html.

29   D. Deutsch, *The fabric of reality: The science of parallel universes and its implications* (New York: Penguin Books, 1997). The book has been highly rated as shown by an excerpt from one review by Professor Paul Davies: "Deutsch ... confronts the deepest questions of existence head on, challenging traditional notions of reality with a new worldview that interweaves physics, biology, computing and philosophy." Thus, the coverage is strongly interdisciplinary in keeping with the present book. This in itself suggests a close connection between all elements of our physical existence and indeed our enduring endeavors.

30   J. R. Brown, *Quest for the Quantum Computer* (New York: Simon & Schuster Touchstone Books, 2001) was previously published as *Minds, machines and the multiverse: the quest for the quantum computer* (New York: Simon & Schuster, 2000).

31   Ibid.

32   S. Lloyd, *Programming the universe: A quantum computer scientist takes on the cosmos* (New York: Knopf, 2006). Lloyd was accompanying Heinz Pagels on Pyramid Peak when the latter fell to his death. Many ideas that Lloyd followed through on had their origins or at least their inspirations with Pagels, his PhD thesis advisor. Lloyd is at the MIT Research Laboratory of Electronics in Cambridge, Massachusetts.

33   R.   F.   Roth,   (2002)   http://paulijungunusmundus.eu/synw/pauli_parapsychology_p1.htm.

34   W. W. Hammerschmidt, *Whitehead's Philosophy of Time* (New York: Russell & Russell, 1975), 100. This is a reissue of the 1947 first edition.

35   Ibid.

36   Ibid.

37   R. Feist, "Whitehead, God and Relativity," in William Sweet and Richard Feist, eds., *Religion and the challenge of science* (Farnham, UK: Ashgate Publishing 2007).

38   Ibid.

39   Ibid.

40   A brief discussion of Jean-Marie Guyau's philosophy is in J. A. Gunn, *The problem of time* (London: George Allen & Unwin, 1929): 242–44. Guyau's *La Genèse de l'Idée de Temps*, conceived before 1888, provides a source for his perspective on how we approach the future. Whitehead's 1933 *Adventures of Ideas* (Cambridge: Cambridge University Press) is the source for his view of the dynamics of our future.

41   http://www.experienceproject.com/question-answer/Have-You-Ever-Thought-Why-Was-It-Me-That-Was-Born-why-Am-I-Me-And-What-If-It-Was-Someone-Else-That-Was-Born-Instead/1261065. In an interview

with Dr. Jon Klimo in US Psychtronics Association Newsletter 3(1) 2017, 27, Klimo revealed that his interest in unexplainable phenomena can be traced back to when he was a boy "standing alone beneath a clear night sky and looking up at the stars and feeling full of awe and wonder." This is exactly what happened to me in about 1945, and I wonder how many others.

42   E. D. Mitchell, *Psychic exploration: A challenge for science* (Florida: Capricorn Books, 1976); foreword by Dr. Gerald Feinberg; twenty-nine contributors. Text refers to page 32 of the introduction to Mitchell's book.

43   Ibid.

44   S. Blackmore, *What we believe but cannot prove*, J. Brockman, ed. (New York: Harper Perennial, 2006). Another reference worth checking out is D. Wegner, *The Illusion of Conscious Freewill* (Cambridge: MIT Press, 2002). I independently discovered this perceived human condition of no real free will in about 2007, and later, there had to be some caveats applied (e.g., the Dunne effect).

45   P. K. Westen, "Getting the fly out of the bottle: the false problem of free will and determinism," in *Buffalo Criminal Law Review* 8 (2005), 101–54.

46   B. Haisch, *The purpose-guided universe: Believing in Einstein, Darwin, and God* (Franklin Lakes, NJ: New Page Books, 2010), 222. See also www.youtube.com/watch?v=fpS0bHmVpz0. Haisch extensively discusses the perennial philosophy (PP) not mentioned in an earlier book. The PP has ancient roots and explains to those who don't already know that we are not "just complex organic machines" or in Hawking's classification as chemical scum but are individually connected to a nonmaterial, immortal soul.

47   Ibid.

48   F. A. Wolf, *The Spiritual Universe* (Needham, MA: Moment Point Press, 1999). Wolf is a physicist with a spiritual and philosophical inclination who once tried to show that the soul works by quantum mechanical principles. He must have realized later that the soul must be governed by a physics that is outside the walls of current science.

49   M. Kafatos and R. Nadeau, *The Conscious Universe* (New York: Springer-Verlag, 1990). D. Radin, *The conscious universe* (San Francisco: Harper Edge, 1997). This one explores the link between mind and matter and involves statistical data analyses.

50   F. W. H. Myers, *Human Personality and its Survival of Bodily Death* (Charlottesville, VA: Russell Targ Edition, Hampton Roads, 2001). The first edition was published in 1903; this book is a reprint of a later edition. The use of the word *consciousness* should accompany *personality* in the above title.

51 http://en.wikipedia.org/wiki/As_You_Like_It. This was the correct title in the Template. Also, the fourth line should read, "They have their entrances and their exits" to preserve causality. Imagine watching the forty thousandth production of a Shakespeare play. You are so absorbed in the acting that you think Hamlet is making freewill decisions. The actors have studied the script so well that to be convincing, they are under the self-induced spell that they are acting out the drama with apparent free will. Consider that this is exactly what is happening in real life!

52 www.enotes.com/homework-help, student (nikki33) in Grade 11 asks: What does the quote "All the world's a stage, and all the men and women merely players" mean?

53 College teacher <gbeatty>. For some who may have forgotten this, my antidote to these "revelations" is to stop thinking about free will and just concentrate on acting your life out. That is what your soul came here for.

54 M. Shermer, *In Darwin's shadow: The life and science of Alfred Russel Wallace* (New York: Oxford University Press, 2002). In the book by Arthur Osborn is a quoted clairvoyant story recorded originally by a Professor Bozzano (Signor Ernesto Bozzano), see: A. W. Osborn, *The superphysical* (UK: Frederick Muller, 1974); first edition was published in 1937. It reads on page 93: "Dr. A. Wallace (Professor Ernesto Bozzano's case 108) received a Mrs. Paulet at his house." This clairvoyant said to his twenty-year-old son who was a student of chemistry: "There will be an explosion in your laboratory in February or March and someone will be injured." Another clairvoyant repeated the same prediction on January 20. On March 9, a terrible explosion took place and severely injured one of the students. That story, poorly documented as it is, may have suffered because it must have traveled from Alfred Wallace—who the Dr. A. Wallace must surely be, to Bozanno and then to Professor Charles Richet, MD (1850–1935) in whose book *Thirty Years of Psychical Research* it is found on page 358. Richet received a Nobel Prize in medicine and was involved in aircraft design and testing; he seemed to be a very inquisitive and brilliant polymath. The evidence that Dr. A. Wallace is our man is also contained in the link: https://archive.org/details/jstor-1640908. As expected, there is much about his life's work as a naturalist but nary a word about his spiritual research.

55 Pierre Teilhard de Chardin (1881–1955) was a French Jesuit priest and philosopher who trained primarily in the field of paleontology (a subfield of geology). His well-known book *The Phenomenon of Man* is a demanding read. In it, he wrote most of the spirit-related matter I have covered in this book using the following two sentences: "We are not human beings having a spiritual experience. We are spiritual beings having a human experience."

His well-known Omega Point is found at http://en.wikipedia.org/wiki/Omega_Point. Although a philosophical topic, the interpretation of the Omega Point based on the structure of the serial multiverse is that it occurs either in the Template or if it has overshot the "maximum level of complexity and consciousness," it is in a universe behind it but likely well ahead of the maximum range of human ability to remote view that far ahead.

56   The scale range within the universe from "quantum to cosmos" covers a minimum of fifty-four orders of magnitude and is arguably at least sixty, i.e., 10^{60}. If the multiverse is added, it would be exceedingly greater than this. I have no way of determining the value. However, it is relevant if we were talking about inter-universe space travel.

57   D. Bohm, *Wholeness and the implicate order* (London, UK: Ark Paperbacks, 1983). It was first published in 1980 by Routledge and Kegan Paul. The amazing thing is that you get his explicit and revealing idea about time on just two pages—211–212—of a 213-page book.

58   Ibid.

59   J. W. N. Sullivan, "Dreaming of the future," in the *Times Literary Supplement* 1339 (September 29, 1927), 659. Sullivan was a mathematician and wrote a popular book on relativity. He was obviously not convinced by Dunne's geometrical model that was presented, only politely saying that it "is not quite satisfactory." This is a vast understatement. Elsewhere, Sullivan said that Dunne's model is "obscure" and "when he [writes] of an infinite number of dimensions 'at right angles' it is not at all clear what he means." The STZM model shows that Dunne must have been referring to the vertical lines, but they are dead, and the "infinite dimensions" are really the dots—or universes—arranged along the upper line. Sullivan seems to be one of the first to take exception to Dunne's so-called "serial time." "Undoubtedly," he said, "the most important part of his book is the record of experience." This refers to the accounts of dreams (and a few visions). The reviewer wrote, "Since dreams of this peculiar character are not [unique] to Mr. Dunne, it may be, as he argues, that they are normal. But if they are normal, how is it that so stupendous a fact has not become generally *known?*" Dunne, and the reviewer, had not checked the literature nor made a trip to the Scottish highlands to interview the people there endowed with second sight. However, Dunne's dream recording protocol, which I began several years ago, paid off when I successfully logged a strong hint of the impending Bow River flood that I mentioned earlier in this chapter.

60   Ibid.

61   Sir William Matthew Flinders Petrie, FRS (1853–1942) was an English archaeologist largely educated by his parents and tutors; he also had an

inborn ability to investigate thoroughly. He became a specialist in Egyptian archaeology and was a meticulous excavator. He wrote several books. The *TLS* reference is 1340 (October 6, 1927), 694. He would have been seventy-four at the time and living, retired, in Jerusalem. The extended Bromby family, from which his cousin came, were typically well educated with university degrees. The now rare 1729 book by William Dampier (*Voyages*, volume iii, p. 290) reprints Wafer's account of the predictions of the Darien native pawawers.

62 Ibid.

63 This quote is also found in Dunne's third edition of his book (1973 reprint) in an "Extract from a Note on the Second Edition," page viii; they differ only slightly.

64 R. L. Mégroz, "Dreams," in the *Times Literary Supplement*, 1341 (October 13, 1927), 715. His obscure book on dreams was reprinted recently: R.L. Mégroz, "*The Dream World: A survey of the History and Mystery of Dreams*", Reprinted by LLC US (2013).

65 N. C. Flammarion, *L'inconnu The Unknown*, English edition (London: Fisher & Unwin, 1900). It contains a selection of his vast collection of anecdotal accounts many of which deal with precognition. Flammarion apparently led a balanced life between studying the unknown and astronomy. His last book published a year before his death was *Haunted houses* (1924). I used to have one of these, and it provided me with a firsthand experience of being in the company of spirits.

66 J. C. Van Pelt, "In quest of experiential anomalies," *Edge Science*, 15(14), (2013).

67 Ibid.

68 N. C. Flammarion, *L'inconnu The Unknown*, English edition (London: Fisher & Unwin, 1900). It contains a selection of his vast collection of anecdotal accounts many of which deal with precognition. Flammarion apparently led a balanced life between studying the unknown and astronomy. His last book published a year before his death was *Haunted houses* (1924). I used to have one of these, and it provided me with a firsthand experience of being in the company of spirits.

69 R. Penrose, *Cycles of time: An extraordinary new view of the universe* (London: The Bodley Head Press, 2010). The quote is from the preface.

70 D. Bohm, *Wholeness and the implicate order* (London, UK: Ark Paperbacks, 1983). It was first published in 1980 by Routledge and Kegan Paul. The amazing thing is that you get his explicit and revealing idea about time on just two pages—211–212—of a 213-page book.

71 J. Barbour, *The nature of time*, fqxi.org/essay (2008); see also note 3.

72 J. Barbour, *The end of time* (Oxford, UK: Oxford University Press, 2000). He said, "The block universe picture is in fact close" to his theory. Thus, it is an incomplete treatment without some form of built-in but artificial motion implied. This is supplied in the serial multiverse model, which is associated with Newtonian mechanics. The Now concept is explainable only in a Newtonian setting rather than being entangled in a space-time setting. I have quoted mainly from page 142 of Barbour's book. He has evidently not stressed Whitehead's ideas on events as being more fundamental than time.

73 J. Barbour, *The nature of time*, fqxi.org/essay (2008); see also note 3.

74 In one respect, this is a positive feature as it is well known that equations drive away many readers. However, I could not do otherwise; this book is not a dumbed-down version of a much more grandiose version. It is essentially the kernel of what I have to say about time and the multiverse after working on it for a decade, which included extensive rants, rests, reflections, and revisions while trying to accomplish other conventional research as I found time.

75 J. Brockman, ed., *This idea must die* (New York: Harper Perennial, 2015).

76 The origin of our universe—be it the Big Bang or something else less violent—would be the first *event* in the history of our universe. I have pointed out elsewhere that human ingenuity using theoretical means and utilizing the results of astronomical observations has consistently produced an age for the universe of between 10 and 15 billion years (in type 2 time).

CHAPTER 6

Conclusions

The following list of essays is a representative selection of the more important conclusions that may be drawn from the material in this book. They are developed mainly from the results and discussions in chapter 5. While I wrote this final chapter, the Internet became very active on matters of time and of multiverses. As a result, some of these concluding essays contain new material.

A Brief Overview of the Multiverse Blueprint and Defining the Two Types of Time

The question that originally led to my decision to mount an investigation resulting in this book involved how to rationalize the extraordinary but not uncommon observation of people being able to see (in dreams, visions or some other channel) future events that subsequently actualize.[1] Most scientific people would shrink from this task just as they do with astrology. It was soon obvious to me that understanding *time* was going to be crucial. This insight appeared in the 1929 book *Science and the unseen world* by mathematician-physicist Arthur Eddington, but John Dunne had already made the first courageous step in that direction with *An Experiment with Time* published two years before.

My approach was to start from basic concepts in philosophy and physics, build a model of cosmology incorporating time and see if precognition was compatible with the model and possibly a verification of it. After encountering the relativistic concept of block time, I realized the task involved developing it into a realistic dynamic state. By asking the right questions and applying inductive and deductive logic, I eventually established the Blueprint for a multiverse. Such a process may not be of much interest to those who make a living by writing mathematical equations but should not substantially raise the ire of those who carefully use the scientific method in a qualitative context.

All that was required was the most basic forms of geometry, the fundamental and most venerable member of mathematics. I embraced Whitehead's classification that formal mathematics belonged within the field of logic. This was the essential vehicle for developing the geometry. Trigonometrical expressions were unnecessary.

The initial stages took place without the presence of reference axes, which are normally entered before developing a graphic. Such a procedure is very unusual for beginning a graphic. The steps in arriving at the Blueprint (figure 2.1A) are illustrated in chapter 2.

When the penultimate diagram was reached, only then could orthogonal axes be added as the last step. The process involved deductive logic. This meant that I had worked in a sense backward—like reverse engineering. At that stage, it became obvious that I was dealing with two types of time. The first one is fundamental to the active functioning of the multiverse, but it is a *timing system* existing for a purpose different from that of our familiar clock time. The former one is fundamental to the multiverse— evidently traceable to an oscillator (acting as a timing device) and had to have been in existence before our universe started up. I link this to the metaphysical time of Feist, type-1 time by me, and what Whitehead termed nonformal time in the second version. The other, most-familiar time was called formal time (by Whitehead), physical time (by Feist), and type-2 time by me.

The horizontal axis in the Blueprint has multiple labels: space— events (motion)—clock time. The dynamic events line (*DEL*) is the upper line which represents the current physical multiverse. All the lines below it are trajectories of past events; the vertical lines imply times past, but they must be fundamentally considered as evolutionary lines that trace events; thus, they have only direction—upward. The label Direction of Evolution is thus shown with an up arrow. Each line culminates at a dot on the *DEL*. All such dots order the centers of individual universes in the multiverse. It is drawn straight only for geometrical convenience; in the special Cosmological model it will be shown to be curved.

I have so far managed to avoid using the word *time* in support of a much earlier deduction that time is not a fundamental physical variable but rather a scalar value used in quantifying movements (or change) that bring about events. This can be provided only by a human built device for gauging, for example, the rate of change of the size, shape, rotation, position and temperature of any physical object in the universe.

In figure 2.1A, the outside vertical axis is labeled Cosmic Multiverse Timing System and operates in a relative way. Here is where the metaphysics really comes into use. The timing system's origin is in a precise oscillator; its manifestation is in the important Now moment, or just the Now. Our mind has great difficulty thinking about the mercurial nature of the Now as pointed out by Einstein (chapter 5). The time pips from the Cosmic oscillator of frequency f where $f = 1/\Delta t$, signal the occurrence of the GEs, the next forward movement of all the universes in the torus manifold, and the rearward copy procedure of each Now. This seemingly impossible process is supposed to be simultaneous or very nearly so. (This was described graphically in the last chapter using figure 5.1b.)

The magnitude of this frequency, or alternatively, the size of the slice of duration Δt, is beyond my reach. In any case, the lowest the frequency it can be is such that it must not cause sensible judder on the retina of the most highly sensitive human eye.

This timing system serves as the pacemaker of the multiverse. Yet it has a print-through into our organic domain. If f could be identified and measured that in principle could lead to verifying the metaphysical hypothesis I have offered here.

Another domain where it might be detected is in the process of fast crystal growth in a saturated solution. Crystal growth should be subject to the same finite process of change.

Time Zones and Precognition

In the STZM cosmological model (figure 2.1B), the timing system maintains and coordinates the multiverse's dynamic character. The continual production of GEs at one location gives rise to a form of serial time that is manifested in the time zones, one per universe. This is not related to the serial time postulated by John Dunne that has been shown to be erroneous. The newly defined serial time is a requirement along with the existence of serial universes for setting up the conditions that allow the phenomenon of precognition to occur. It further requires understanding the existence and functioning of consciousness, the process of dreaming, and the totally unknown process of information transfer from one universe to another over immense distances that is required for precognition to work. This is equivalent to invoking full-scale action at a distance, which is a property of the current version of quantum mechanics.

Furthermore, because the treatment of time in quantum mechanics has yet to be fully understood and as we are using a Newtonian space-time system with no gauge length based on the speed of light, there is no basis for placing limits on the speed of objects or information transfer in the multiverse; Einstein's theory of relativity is inoperative here. Indeed, even the solution of his equations for the cosmological case for our universe is Newtonian in nature. This condition must by inference be therefore the same in all other universes.

The setting up of the time zones in the universes occurs automatically. At each GE, *that* universe's clock starts running. But it is not yet a physical clock, only a potential clock. This start from zero applies whether it is an (undefinable) explosive beginning or a finite dimensioned piece of mass that undergoes a transformation that results in a slow but accelerating expansion as seen in figure 2.1B and in the sixth essay ahead.

When humans invented common clock time or type-2 time, they conveniently provided a basis for measuring the Now and, indirectly, the duration of the time shift Δt between adjacent universes. Necessarily, when thinking along these lines, we have to be unfazed by the huge scale of the multiverse; consciousness seems to have no restrictions, but the logical mind has trouble dealing with it. Thinking non-locally as having no limits will help reduce this difficulty. It is here that I may have provided a reason for keeping mind and consciousness separate but still allowing an interaction. However, these are two areas which are still evolving in the minds of Man.

The Elusive Now and How It Might Work

In figure 5.1, I attempted to investigate the dynamics of the Now moment because it naturally presented itself when I applied the well-known concepts of numerical computing to operations in the Template. For those readers who are not aware, variables in a continuum equation must be discretized and then rewritten in computer code, which allows a solution of the original equation to be made by stepping (forward) one quantity in the equation. This could be time, or distance being marched around the loop of the physical multiverse (figure 2.1B). Einstein correctly considered that the Now has great significance to humanity, but he concluded in about 1950 that it was just beyond our capability to deal with. Here, I am suggesting a way of thinking beyond that situation as Bohm was trying to do possibly because of discussions with Einstein at Princeton.

The Now is taken as representing a finite moment of reality constituting part of all the scenes (events) taking place simultaneously in a particular universe. Originally computed in the Template this huge file is copied (via the COPI procedure) from there and passed along the line of universes behind it. This direction is opposite to the movement of the actual universes in the torus. Each step in that process has a duration Δt. In figure 5.1, I indicated that the Now may be of duration slightly less than Δt, if simultaneity is denied. This procedure differs considerably from what Whitehead suggested in about 1933 when he speculated that the "future is inserted in the crannies of the present." This can now be seen as a highly imaginative metaphor.

I will attempt to walk readers through an imagined animation of a part of the Cosmological model (figure 5.1b). I will deal with only three universes through which the copy procedure passes in succession. Focus on the middle one, which is typical of all copied universes. There will be a regular synchronized time pip (TP) of short duration in the three universes. Follow the COPI instructions (in the caption) that updates the Now file in the middle universe. The current Now there was just previously present in the universe to its right. As the new Now in the middle universe, it will remain for a duration of $\leq \Delta t$. Practically simultaneously, the universes in front of and behind the middle universe will get their new Nows; the one in front from the universe just ahead of it (to the right), and the one behind from the middle universe. Simultaneously, each universe shifts forward an increment, Δx say, to gradually advance its way toward the position of the universe in front of it as seen in the grid system of figure 5.1a.

This process involves all the universes from the Template, all the way along to the proto-universe right after the last GE. The Template has its own calculate-and-paste-in routines (CAPI) based on the same physics as applies in this universe. This does not mean that they are identical to the physics as is currently described and used.

The Structure of the STZM: How Does It Measure Up?

How many multiverse hypotheses are there? Michael Hanlon[2] says there were nine in 2012. Since then, another two can be added: the recent many interacting worlds (MIW) hypothesis[3] and now the serial time-zoned multiverse (STZM) geometrical model of this book, which brings the count to eleven. The earlier hypotheses seem more to represent works in progress. None of the schemes is couched in the same terms as the STZM model. In the case of the MIW model, the universes appear to be in terms of physical properties compatible with the STZM model universes, but as I pointed out in chapter 5, the MIW scheme (which has a formal quantum mechanics component) is not completely transparent. The STZM doesn't have a formal quantum mechanics component, but one can be inferred by recognizing the need to have a Cosmic quantum computer.

How many physically explicit models have been proposed for the structure of a multiverse? I exclude here some hypothetical configurations that defy three-dimensional graphics and reality. I refer to a verifiable three-dimensional (multiple space domain) model that provides useful information. A three-dimensional space packed with bubble universes all in contact offers no obvious information that will lead to cleanly solving even one of the unexplained natural phenomena or the contrived paradoxes I have dealt with in this book.

It is a well-established principle in the scientific method that a viable theory must be testable. I have shown that this is the case for the STZM model provided you can agree that precognition exists. Importantly, during the development of it, the mental block attached to the concept of block time was revealed. It was shown to be a means of stepping forward in the task of taking the confusion out of the troublesome word *time*. The STZM immediately opened a window on the mystery of precognition and can be used in any of the Friedman-Einstein cosmic evolutions to reveal fallacies in certain concepts found in physics such as the bizarre cases of the bouncing universe, a-causality (or retrocausality), and the grandfather paradox. The slow accelerating expansion at the start of the GE as shown in figure 2.1B may suggest why the cosmic microwave background temperature of the universe is so uniform. According to Andrei Linde, this level of uniformity is hardly achievable with a big bang. It would make the radical inflationary hypothesis unnecessary, a move that would appeal to Sir Roger Penrose. In addition, I am not the only one to postulate a slow, low-capacity universe start-up.[4]

Though I cannot second guess the structure of some of the other proposed multiverses, I was somewhat encouraged to read in a book by M. Chown[5] that first, M. Tegmark thought it unlikely that there would turn out to be just one universe capable of supporting "self-aware substructures." These substructures are assumed to be Hawking's "chemical scum," an excessively demeaning category to which we are all supposed to belong. Second, Tegmark thought that "there is an archipelago of universes." This is rather suggestive of a serial multiverse arc, is it not? Third, he maintained, "The most common universes will be the ones with equations of physics that are very close to ours." This is where his thinking diverges from the STZM and the MIW models. In those schemes, the physics in each universe is identical; otherwise, the scheme fails.

The STZM exhibits a quality of tight geometrical coherence obviously responsible for its consistent ability to address many problems that have appeared as puzzles to philosophers, physicists, and dual versions of those alike. The reader should know that I already had the manuscript for this book edited when the MIW article appeared in late 2014. Thus delays were caused by extensive revisions being necessary throughout the book.

At the beginning of the multiverse project, I had speculatively targeted quantum mechanics (QM) as being the driver of the mechanisms acting in the multiverse simply because the COPI routine required that everything in the universe—from fundamental particles to the largest structures in the cosmos—had to be regularly transferred in one huge file from each universe to the universe following it. While my approach doesn't and cannot explicitly include QM, some association with MIW—which is based on QM—would seem to imply that QM is present in STZM. In interviews, commentators[6] listed several essential features of the MIW multiverse. While the MIW formulation is unclear on the dynamic nature of its universes and on the actual large-scale structure (on the scale of figure 2.1B), I was able to identify quite a number of features of it that coincided with ones in the STZM model.

I list below seven features that constitute a bundle of conditions and requirements showing that the main features of the STZM model are broadly similar to those in the MIW formulation. My inputs to the list occur as comments added in brackets. I will reiterate that I had finished my synthesis long in advance of being alerted to the publication of the MIW formulation.

While the first two points read like initial conditions, the third (edited), fourth, and fifth points seem to have been emerged from the mathematics. The requirements and conditions of the MIW theory accompanied by my responses in brackets are these.

1. There is a "gigantic ... unknown number of worlds" that are "equated with universes."[7] [These statements all apply to the STZM.]

2. There are a finite number of universes[8] in the model. [Such is the case in the looped STZM.]

3. "Some of these worlds are almost identical to ours ... but ... most (worlds) are very different."[9] [These conditions accord with the STZM if I correctly assume it to be an evolutionary effect. Recall that all the universes are strung out around the torus as shown in figure 2.1B. Every universe is currently active and evolving just as this one is, and as a result, we can state that early universes are very different in size and content from much later ones. It is obvious that by far the number of universes would be "very different" from now because of our knowledge of the history of our own universe.]

4. All the universes in the MIW model interact or are linked as a population. [All the universes in the STZM model are arranged serially and are continually interacting as a result of the COPI procedure, but they are also required to be interacting in another way to account for the special evidence presented by the established presence of communication between universes. This occurs over a wide band of type-2 time—equivalent to a long string of events into the future available to a percipient whose consciousness is connecting with a specific future event. This implies that a continual flow of digital information is involved.]

5. "All the worlds are equally real, existing continuously through time with precisely defined properties."[10] [This applies in the STZM except that the use of the word *continuously* through time in the MIW formulation—which is due to the use of analytical equations—is not the same as the numerically driven formulation in the STZM model. This requires the use of the word *continually*.]

6. Quantum phenomena arise from "a universal force of repulsion between 'nearby' worlds."[11] [While there is no equivalent result or condition to this in the STZM model, the STZM benefits from its adoption as we saw in chapter 5, which I advantageously revised. The metaphysically branded result indicates that the outer surfaces of universes have sign parity in their charges—and hence they repel but just so far due to gravity, which acts in the opposite direction. The result is that a small gap is maintained between universes. By postulating an ERB (worm tube) to exist in this gap supports the conditions of point 4 above as covered in chapter 5.]

7. Both models can exist within a general Newtonian space-time domain. In the case of STZM, a neo-Newtonian[12] description might better fit the properties of the multiverse Blueprint.

This last point, not directly related to the first six points, hinges on a discussion of *metaphysical time*, a term used by Richard Feist (chapter 5 note 25) linking to a reanalysis of some material out of Whitehead's book *Process and Reality* previously made by "some commentators." Feist concluded that it was "our space-time cosmic" domain that grew in Whitehead's "extensive continuum," which must be in the torus manifold but outside the universes. I have taken this to be equivalent to what is seen in figure 2.1B (visualized in motion). Thus, while Whitehead and Feist did not discover the multiverse, they touched on an important geometrical element of it and identified a metaphysical type of time, which seems to correspond to my type-1 time and Whitehead's second version of nonformal time though I do not know how he formulated it.

If two theories derived from different methods show common characteristics, that gives credence to features of both theories. In the case of the MIW model, the starting point was the theoretical domain of quantum mechanics, whereas the STZM model was developed from the concept of the moment of time called the Now, which was linked to a super Cosmic domain size. These combined results should put paid to the rants of multiverse scheme critics, one of whom said as of June 2013 that speculating about multiverses "cannot lead to any real scientific progress because we cannot confirm or falsify any hypothesis about universes causally disconnected from our own."[13] In chapters 3 and 5, this statement was found to be untenable. Moreover, the hitherto philosophical concept of eternality (eternalism) is shown explicitly in the STZM.

Illusions: a Bouncing Universe, Retrocausality, and Even Free Will

There exists a strange conceptual model for the evolution and extension in cosmic time, of a single recycling universe that has been published in *Scientific American*. A graphic artist (assumed to be following instructions from a physicist or a graphics editor) rendered the path of the center of the universe at successive stages of its depicted bouncing[14] flight over several cycles. These involve expansion (from zero), contraction (to zero), and then the rebirth of a new universe experiencing the same or similar geometry as the earlier cosmic cycle. It originated from simply extending in cosmic time Alexander Friedman's result for the first cycle—the so-called closed-universe solution. Textbooks label the time t as cosmic time. The solution for the transient scale factor variation (which can be translated into the radius R of the universe) is given a mathematical description[15] showing there are singularities at the start and finish of a particular cycle.

I next summarize the graphics demonstrating how the term *bouncing universe* came about and show how it was an illusion created when the original graphic was rendered into an incorrect form as shown in figure 6.1a.

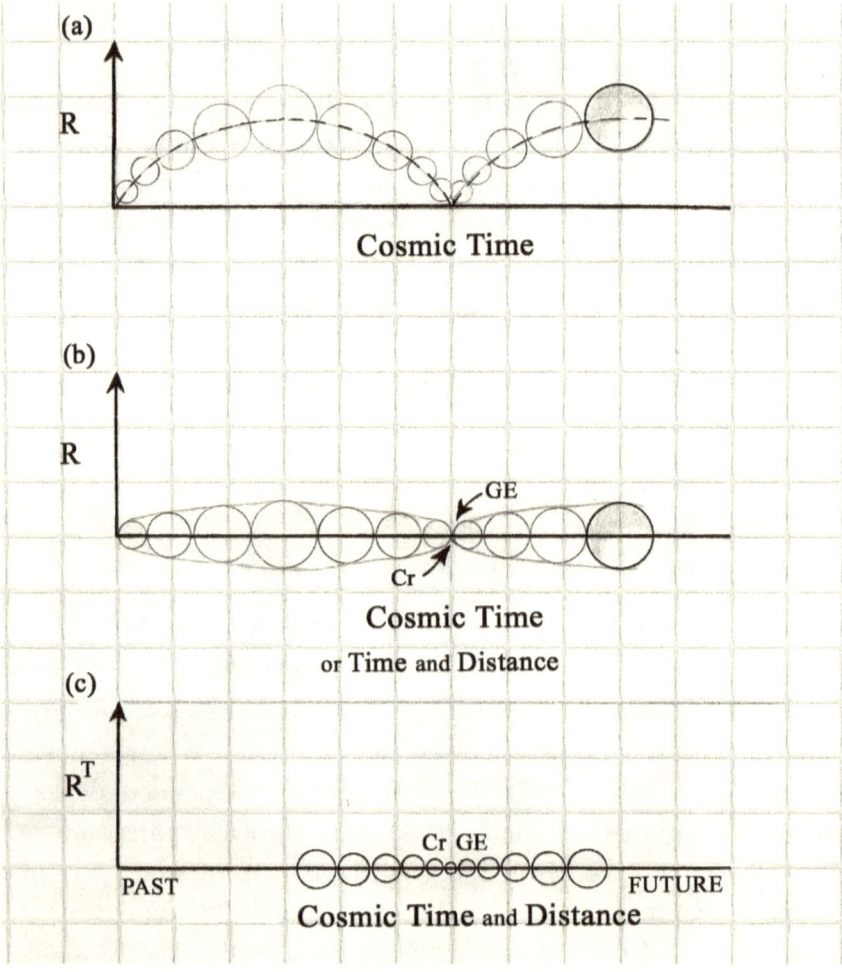

Fig 6.1(a) Dashed curves represent the changing value of **R** (a scale factor that can be taken as the radius of the universe) versus cosmic (which at the universe's outer surface merges with Cosmic time), for one and a half cycles of a closed universe (see Chapter 2). The mistaken, view of a *single* universe in bouncing mode is shown following the dashed curves–giving the *impression* of a bounce. **(b)** The correct representation of the variation of **R ≥ 0** *above* the horizontal axis after a 1/2 scale reduction from (a). Below the axis is the mirror image. Here, the full circles are *separate universes* with diameters

that match the upper transitory set in (a). Thus, as shown in Chapter 2, the cosmic time history of a *single* universe sets the pattern for a *complete* multiverse as represented by the group of 7 circles; the next 4 circles are half of a repeat cycle. **Cr** is the position of the Crunch and **GE** is the Genesis Event immediately following it. The limitations of the graphics process results in distortion in the nodal regions and not being an attempt to represent early universe inflation and late universe deflation. Panel (**c**) is a zoom-in of the **Cr** & **GE** region (R^T indicates another scale change) showing the smooth transition as universe 'material' moves from *left* to *right*. Use of tensed terms: **PAST** & **FUTURE** avoid the use of an arrow. The arrangement of spheres is the 180°*reverse* sense of what is seen in figure 2.1B, where universes travel around the torus *clockwise*. The arrangement in (**c**) can be converted to agree with Figure 2.1B by reversing the movement direction of the spheres, and switching Cr & GE and **PAST** & **FUTURE** but this produces a diagram which is in the opposite sense to diagrams found in cosmology textbooks.

The use of Cosmic time in the last three diagrams means that it should be taken as the time that is outside the universes in the domain of the multiverse timing system; it corresponds to my type-1 time.

In retrospect there seems to be two categories of illusions. The first category is human generated—those produced by the misunderstanding of the theoretical underpinnings of published results. The second category includes those produced by the particular way we look at our existence and our surroundings.

In the first category is the one we have just dealt with, the so-called bouncing universe; it is the grandest one in the group. It is closely followed by the false problem of retrocausality and then the conviction by some people that there is only one universe. After all, why settle for billions of universes, they say, when just one will be enough? Dunne was a notable victim of this illusion as was Bernhard Haisch but for different reasons. There may be other beliefs in this category that you are aware of.

In the second category and perhaps the most significant to us overall is the illusion of free will. A minority of people can figure this out for themselves, and even fewer tell others about it. A feeling of free will comes naturally as part of the development of the physical and social habitat offered to organic life in a small part of each universe. The free-will feeling is very pervasive as it may be inferred that every human life is supposed (by the architects of the multiverse) to be lived as near as possible to what it is in the Template. Yet, it may be

exposed as an illusion. It clearly has ramifications in the area of ultimate responsibility of individuals in criminal trial cases as discussed
in chapter 5.

The Smooth Cr-GE Transition: More Metaphysics

Fig 6.1c shows a close-up of the pinched-down manifold envelope
where the second cycle begins at $\theta = 0°$ (see figure 2.1B). The manifold
(in figure 6.1c) is geometrically compatible with figure 2.1B except that
the direction of movement of universes is displayed as reversed, which
is a result of the particular movement conventions used in the original
graphics. At $\theta = 0°$, the Crunch position, it is a complete mystery as
to what has happened to a once-burgeoning and marvelous universe
filled with visible mass, dark mass and dark energy.

Recall that G. Lemaître believed there was a good case for the existence of a big bang at the beginning of the universe, but that was based
on the Friedman-Einstein-Lemaître solution for the evolution of the
size of the universe. The existence of a mathematical singularity at the
beginning and end of the cycloidal curve cannot be taken seriously for
indicating an explosive beginning. Besides, no part of the rest of the
curve can be matched by modern astrophysical measurements. Even
Einstein was uncomfortable with Friedman's cosmological solution;
that surfaced later when he said (in one of his famous quotes) that even
if the mathematics is correct, it does not necessarily correspond with
reality. Being a Jesuit, Lemaître may have been predisposed towards
spectacular creative beginnings but at least he did postulate a cosmic
primeval egg, rather than hold to mathematical zero at the beginning.

Over three decades ago, Alan Guth (a prominent proponent of
the universe inflationary theory) and a collaborator claimed that a
bounce couldn't occur, but no mention was made of the graphic illusion illustrated in the previous essay. By making some assumptions,
this no-bounce result was based on three of their own results obtained
from quantum theory. In broad terms, this would seem to support
figure 6.1c though I am unable to elaborate on this.

Recently, the quest for knowing what happens at $\theta = 0°$ has been picked up again by Bojowald[16] and others who are attempting to use modified quantum theory to allow a smooth splicing between the end of one cycle and the start of the next. So far, no breakthrough has been announced, but regardless, the research has been presented in the science progress media. The conclusion is that my apparent science fiction excursions in chapter 2 as to the nature of the finite transition at $\theta = 0°$ do not seem to be quite as far-fetched as might have appeared when I advanced them. At least the MIW model has supported the STZM model because of the specifications provided by a multiverse containing a finite number of universes. The focus on a smooth transition will take the bounce out of any attempts to model this section of the multiverse manifold.

A Good Use for Wormholes

This has the nature of an excursion into fringe physics[17] where equations do exist. However, the issue is whether solutions have been interpreted correctly. The basic idea behind the scheme is that the near-contact surfaces of individual universes are locations possibly suited for the existence of a wormhole-like geometrical feature; it could represent a portal connecting one universe to another enabling the inter-universe communication that we have seen is necessary for the proper functioning of the STZM. It is rather extraordinary that wormholes, which originated from the Einstein-Rosen Bridge concept, seem to have been taken more than half seriously by some mainstream theoretical physicists who are not short of creative ideas themselves. John Wheeler was an example of one of these techno, avant-garde physicists.

Many different types of wormholes are displayed on Internet sites, but that invites cherry-picking—a dreadful habit that occasionally creeps into science research. However, I found a case[18] in which Stephen Hawking apparently did or was helped into doing some cherry-picking. There is a diagram in that article engineered to fit the bill for what I had in mind while on one of my Internet *reconnoitres*; it may still be found on other Internet sites as well. This is particularly fortunate because the authors of the article in note 3 stated that there would be a repulsion between universes and yet a need for communication between them. This repulsion—moderated by gravitational attraction—should provide a slender gap between universe boundaries. This suggested the ERBs depicted in figure 5.1.

A Perceived Paradox Identified as a Case of Soft Free Will

A paradox usually suggests that somewhere, an erroneous assumption has been made. Such seems to be the case in the grandfather paradox. In chapter 5, I used Dunne[19] as my mind experiment protagonist, but I could just as well have projected this new thought experiment on myself except that I did not personally experience any future event dreaming to get me started on this book. However, I subsequently experienced a future-event dream in 2013 and captured it only because I followed the protocol Dunne had laid out in his 1927 book.

Dunne would not have thought of this experiment because of the belief that his scheme did not involve a multiverse, though his figure 8 indicates that he should have reconsidered this option. In what follows, I will avoid using the loose word *copied* and use the more explicit and possible command *cut out and paste in* (COPI).

The thought experiment goes as follows: Dunne (as a physical body but as a soul different from the one that occupied the COPIed body of John Dunne in this universe) lives his life in the Template first as a precocious and crippled boy, then a student at Eton, then a soldier in the second Boer War, and later as an aviator who designed, built, and flight-tested a series of rickety, underpowered, and potentially self-destructive airplanes. Being in the Template, he could not have had any precognitive dreams; therefore, he would not have had reason for writing a book on the subject and thus nothing to stimulate writing a treatise on the geometrical representation of time. In principle, he might have been equipped to see only past events, but that may not have been sufficient to spur him to write a book about it.

As the subsequent COPI procedures continued back along the universes behind the Template, the multiple John Dunne bodies in those universes (copied bodies with different souls) began to get glimpses of some future event until in one universe, one of them dreamed so much of the future and was faced with enough evidence to be motivated to tackle the problem of explaining this seemingly improbable experience. It involved the nature of time. One day he shouted, "Eureka!" He rolled up his sleeves, sat down at his drafting table formerly used for designing aircraft, and began drawing diagrams for the book *An Experiment with Time*. This book became a new physical entry into the file. It was obtained just by rearranging material (and ideas) already present. Nevertheless, a new event had sprung up.

Eureka events such as the one John Dunne experienced ought to qualify as soft free will. I call it the Dunne effect. I have posited that there might be a critical energy threshold that has to be exceeded to break the grip of determinism. This would placate those who voice frequent complaints that I overstress determinism to the exclusion of free will despite my demonstrating that in principle, free will is incompatible with a copied universe. I stress that determinism applies correctly in the STZM model at the large scale.

It would be instructive to report on some unusual data that came to light after 9/11 and occurred above the local scale. The data of interest were being recorded not long before the infamous 9/11 event. Online, guided, prescience-testing experiments were being routinely run and results were being continually recorded.[20] An unusual phenomenon showed up in several of the results. Instead of remote viewing ahead over short durations and predicting a random, machine-selected picture from the stored experimental collection (to be momentarily displayed on the computer monitor), the volunteer subjects appeared to be describing images that were later judged to be vignettes of the imminent 9/11 event albeit containing some symbolism. Thus, they were remote viewing much further ahead in time being distracted from the experimental task at hand by a larger-scale, high-energy event that was imminent in New York (and related events elsewhere). This phenomenon seems to indicate support for the hypothesis that the energy of a large event is capable of affecting numerous people closely in time. Yet this did not result in a whistle being blown! Do these *intrusions* (a term used by John Dunne) demonstrate the power of determinism at the subconscious level?

The *Titanic* effect (TE) described in detail in chapter 5 is another case in which the energetics of the event (associated with the dying of 1,500 people) may have exceeded a threshold value and led many clairvoyants to perceive holographic-like images of the disaster scene. There are several possible scenarios. Take one of these as a conjecture. A vivid dreamer or visionary sees an impending air travel disaster in progress and links it to future travel plans he or she has recently made. Immediately following this dream experience, the person is faced with a decision whether to act on the dream or ignore it. The percipient may assume that he or she is directly involved. Taking evasive action could save his or her life and possibly those of others. So again, if (ostensibly freewill) action is taken the COPI file would now suddenly contain a new piece of information not present in the Template. How many of these cases are documented in which the TE is proven?

For the firm believer of free will, such a situation will reinforce that conviction. I expect that the intention of the multiverse designers to locally soften the grip of predetermination was to allow free will to appear in limited amounts determined by energy controls and the sensitivity of the percipient. This would tend to make a life appear self-driven and thus make it feel natural and thus more fulfilling. It is interesting to note that dramatic, warning-type dreams seem to occur willy-nilly; they do not occur regularly enough that might allow this type of dream (or more rarely a vision) to proliferate or reveal their mode of occurrence.

Despite the weight of support for predetermination based on the STZM as well as empirical data, a person may (in principle) still be able to experience a type of free will, which I call soft free will, though that is a borrowed term. It denotes a local effect that diminishes rapidly away from the source and with increasing duration after the experience.

A Brief Assessment of the Legacy of J. W. Dunne

Why do I keep returning to John Dunne? Here, I am referring to all the John Dunne's who exist in the multiverse. It is because he (really they) provided me with the means to coin the Dunne effect, which is connected to soft free will, and he (they) beyond the Template probably spent as long as I did working on the problem of how to represent time geometrically. No one else to my knowledge has done this. For me (plural), it was an exercise in detective work to follow his thinking. It took the STZM model, which was developed independently of his work, to enable me to figure out how he had managed to construct the first geometrical model of time. He coined the term *serial time*. Independently, I arrived at a serial time, but it is fundamentally different from his.

Once you understand Dunne's mind-set, you can understand why his second book, *The Serial Universe*, was an immediate failure; it was ignored by influential physicists such as Eddington and Bohm. The last book: *Intrusions?* is a source for finding out about his early life and the revelations about his spiritual experiences. I shouldn't have been surprised if he had confessed to some dream assists when building his model of time, but he couldn't reveal anything about this. In principle, it was possible for him to have dreamed the Pauli dream covered in chapter 4. I identified this as a newly reported archetypal dream, but Jungians may question this.

The idea of serial time came to Dunne as a boy and may have been a hint occurring in a repeating dream. It certainly would seem to have been a promising hint. But if so, he misinterpreted the dream. In *The Serial Universe*, Dunne devised a type of serial time using cartoons that depended upon the existence of many more (infinite) dimensions without specifying where they were.

Dunne's mistake can be identified in figure 4.2, which is a recti-
fied cutout of his figure 8 in the third edition of his 1927 book. There,
the vertical lines can reasonably be matched only with my evolution
lines (as in figure 4.1b). He fixed his diagonal line O'O. This would
have allowed him to have the time offset between the vertical lines
that he needed by making them start at the intersections along O'O.
It wasn't until his last book that he stated that O'O sloped at exactly
45°. I proved this for him as it was a fundamental result at one crucial
stage during the building of the Blueprint. Dunne somehow incor-
rectly associated this line with the slope of the cone in the (relativistic)
Minkowski diagram, which also slopes at exactly 45°. But this (and
any other) association with relativity theory is erroneous. He confused
the picture by drawing useless lines below that diagonal. In fact, his
diagram doesn't have any finite boundaries, which were essential to
the establishment of the STZM Blueprint.

Then Dunne recognized that he needed "entities" arranged in
a row along the line GH, which intersects the diagonal line at O.
Dunne had GH going upward as does the dynamic events line in the
Blueprint. These entities coincide with the universes in the Blueprint.
Even with this hint, he still doesn't go along with multiple universes
because as he wrote in the introduction in *An Experiment with Time*,
"Multidimensional worlds ... dating back to the days of the Indian
philosopher Patañjali, have never appealed to me. To introduce a new
dimension as a mere hypothesis ... is the most extravagant proceeding
possible."

Nevertheless, something showed promise. He had identified and
marked on the diagram in figure 8 two types of time. They were t_1
(along the horizontal axis) and t_2 (along the vertical axis), but he (in-
correctly) wrote these times as vectors. These were the first two terms
of the regression series (proliferating serial times). So he then went a
step further by building an ingenious three-dimensional diagram in
which he accommodated a third time, t_3. This was as far as he or anyone else could possibly go in the geometry of three-dimensions. His
mention that these other dimensions lay "at right angles" to his line

GH confused Sullivan, the reviewer of his book's first edition. Dunne had been correct in this because these are the evolutionary lines for each universe seen in my Blueprint and they afford innumerable space domains each containing a time in a unique time zone scheme. These are the time zones of the Blueprint.

Dunne called the vertical lines the substratum. By moving his line GH up over the substratum, Dunne was invoking movement within the "entities" along GH. In principle, this is what I did in the Blueprint. The real, live multiverse of course is contained in the *DEL and in Dunne's line* GH. This represents the actualization of real events occurring simultaneously throughout the multiverse.

This solves Dunne's dilemma with serial time which is where critics of his theory dismissed his model altogether. The two types of time Dunne labeled in his figure 8 correspond relatively to the existence of two types of time shown in the Blueprint. Thus, Dunne's t_2 time might in principle correspond to the Feist-Whitehead metaphysical time or my type-1 time. The questions arise: Did Dunne dream the same dream as Pauli did? Alternatively, did he dream of the Blueprint? It wouldn't necessarily have been the one in this universe as there are copies spread out in many universes. Both these scenarios are possible in principle. Dunne wouldn't have wanted to admit that he was getting help from dreams— just as Franz Schubert didn't want contemporary musicians to know that he was receiving music scores or music melodies from dreams.

Our physical bodies and the lives they live are in a sense immortal in that every life ever lived is being lived simultaneously by dual bodies in a restricted band of universes where *Homo sapiens* occurs in the multiverse. But except for the originals generated in the Template, they are all copied physical bodies. Dunne later published *The New Immortality*, which he claimed was a simplification of his theory of time. He wrote at the end of chapter 1, "In real time … everything which has established its existence remains in existence … forever." This is exactly one of the main deductions made from the STZM. But I cannot see how he establishes this statement based on reading his first book. It is clearly established for the STZM where the scheme is laid out geometrically. What he was unable to say because he had thus far deliberately ignored any serious discussion of our spirit-derived soul was that our live-aboard souls are also truly immortal.

Then appears the statement "Man … is not accorded distinctive treatment; he merely remains with the rest." This statement might otherwise be considered to lie outside the context of our present subject had it not been for my deduction that the whole point of the multiverse is for the benefit of the multitudes of souls who are seeking real human experiences. But then, the reality of life suddenly hits. I think Dunne might have been basically right if he had wrote that man (and this includes animals) are not accorded special help. But this applies in copied universes; I am not sure about the Template. This is because it seems to me that totally weird things like synchronicities, the Pauli effect and serial significantly linked events, might have originated there.

Continuing Research and Connecting the Dots

In case your mind didn't just flash back to the fourth essay in this chapter, here is another thought to ponder. Of the seven listed properties of the MIW quantum mechanical multiverse and the closely matching properties (bar one) of the STZM model, there is no explicit connection to any spiritual content. In chapter 3 a multiverse verification test was made using a verse from the Book of Ecclesiastes. Due to its proximity to theological material that verse and certain others might be taken as being theological. This is not true; the theme concerns *time* and how it is to be interpreted and applied. In any case, theology occupies a level not seen in the STZM. Yet our *spirituality*, (more fundamental than theology), can be directly seen to explain why the STZM was constructed as it actually appears to be.

This was a totally unexpected outcome of building the multiverse. The particular structure of this multiverse, which preserves actualizations of all the events that have ever occurred, fits the philosophical term *eternalism* (a.k.a. *eternality*) exactly, and it can hardly have been brought into existence for any other purpose than for the incarnation of souls. For what else could such an obviously planned multiverse be used? The fact that the operation of the multiverse is beyond our understanding has no bearing on our contemplations of its function.

All we can and should do is to monitor the not-quite normal phenomena that occur in this universe and by extension the multiverse; that can make us feel more comfortable with our total environment. For instance, remote viewing (RV), which can be performed successfully by only a select number of elite people, can be put to great practical use if the results are of sufficiently high fidelity. An unexpected spin-off of RV experiments intended for strategic interrogations in our present space was that some operators could flip into universes in other time zones at the same spatial locations. Imagine being given the coordinates of a location on this planet and being able to describe it. That is just what some RV operators can do.

It is natural for humans to want to figure out mechanisms. In fact, researchers are encouraged and even required to do this in mainstream science. They sometimes do this at the risk of getting the wrong mechanism. Not to worry; someone else will find the correct one. But in the area of research covered in this book, you are likely to be led into metaphysics. As I mentioned in earlier chapters of this book, I have generally attempted to avoid getting involved with mechanisms and tried to concentrate on the pathways for information transfer that are offered by the explicit and inherently simple structure of the multiverse scheme presented here.

An interesting statement made two decades ago by Dean Radin impinges on a mechanism that connects with my thinking about precognition: "Another way to investigate precognition is by exploring the possibility that the mind is in contact with its own future state."[21] Two points could be made here.

1. The idea that we may be "in contact" with our "own future state" is in keeping with the perceived functioning of the STZM. You and I are copies of our simultaneously existing

future *physical bodies* but which contain *different souls*. That makes the dual body unique. The question is: Is the communication between the souls?

2. It now appears that it is not so much the possibility of this happening but rather in which direction the information transfer is. In the case of the COPI routine, I have established that information must be transmitted counter to the universe's travel direction in the torus manifold. In typical precognitive dreaming in which dreams seem to appear willy-nilly (unsolicited), can one assume that information is arriving via this same counter direction? Alternatively, is it obtained by the dreamer's consciousness (shared by the soul), which is known to be capable of what amounts to inter-universe (trans timezone) travel without the body?

Recall that in directed RV (future mode), the viewer is actively seeking the information; that is, the consciousness (awareness plus mind) of the viewer seems to be able to obtain the information, even classified information, free—without a credit card. Thus, we see that the STZM model has limited capabilities; it can provide only the pathways for assumed information transfer. The mechanisms are largely left up to the ingenuity of the enquirer to figure out with little more than some help from quantum mechanics such as the principle of nonlocality.

Due to an interesting observation of J. D. Barrow[22], I can now end this essay on a philosophical spin. He maintained, "The flaws of Nature are as important as the laws of Nature for our understanding of true reality." Is RV of the future (precognition) an example of a possible flaw, a chink in the defenses of the monster multiverse itself? The number of people who can produce very convincing cases of precognition (including future RV) is judged to be quite low (1 percent or less for precogitory dreams and premonitions) according to independent estimates. I think this value may be unreasonably low caused by too-stringent conditions set by some investigators and not enough people keeping dream logs.

Some Bottom Lines Pulled Together

Because there are a large number of quite different concepts being manipulated throughout this book, it became a challenge for me to link them together in some tangible way and write down a tangible explanation for them all and even for an editor to edit it. The manuscript seems to exhibit *entanglement*, a term borrowed from quantum mechanics. I have occasionally found it necessary to plead for allowing the presence of some repetitious material to serve as a memory refresher for the author as well as reader. To complicate matters new material kept coming online as fast as I expanded the manuscript, for instance, the sudden appearance in late 2014 of a new multiverse theory coupled to quantum mechanics.

The serial time-zoned multiverse (STZM) cosmological model presented here was a particular interpretation of the Blueprint. At that stage, only one of three universe scenarios was available for me to choose from. The scenarios coming from Friedman's solutions of Einstein's simplified field equations included two solutions termed *open* because they expanded infinitely. One solution was termed *closed* because it just appeared, grew, and then shrank to zero again. Mimicking the history of that one universe in the multiverse (figure 2.1B) required that it contain a finite number of universes but the actual number is not known.

One of the specifications of the MIW multiverse was that it too contained a finite number of universes but the exact number was not known. In addition, a multi-properties cross-check on it against the STZM model was available. However, as stated in chapter 5 and in essay number 4, though the elements of the MIW cosmology have correspondence with the STZM cosmology, I do not understand the connections to the quantum mechanical scheme.

To further emphasize our myopic view of time (as in continually consulting your watch), I cannot resist rewriting one of the MIW properties. Number 5 reads, "All the worlds are equally real, existing continuously through time with precisely defined properties." In the numerical computing system, that would be written: "All the worlds are equally real, existing continually through space with precisely defined properties." The mathematically continuous nature of the MIW equations resulted in the use of the word *continuously*. Here, you see again that mathematical formulations do not exactly correspond to reality. This is the point Tesla made. Even Einstein, later in life, espoused metaphorically that God used numerical methods of computation. It was likely this insight that evidently brought Einstein's productivity to an end earlier than expected. It also underscores what Hawking once asked: "What is it that breathes life into the equations?" The short answer is probably that the Cosmic quantum computer performs this function.

Now that some of the secrets of our multiverse seem to have been unveiled to those who have a need for them to be revealed, it is worth reviewing that astounding phenomenon of precognition that made this possible. Once seen as being in principle the easiest of the preternatural phenomena to verify but the hardest to explain, I now see precognition as having a crucial significance in our lives so much so that it should not be associated with several puzzling phenomena in the occult repertoire that are local and truly bizarre effects. Precognition seems most aligned with remote viewing (RV) involving a time shift from the remote viewer's local clock time.

Thus, precognition should be removed from a classification of impossibilities[23]. Imagine a phenomenon that is classified impossible simply because its mechanism cannot be explained in terms of the science of the day.

Several computer scientists believe the universe is operated by a cosmic quantum computer. David Deutsch thinks in terms of a multiverse with a linked network of computers. The STZM platform would cover the large scale Cosmos. Here is the concept of a clockwork-driven universe as Newton's universe was called.

Because the building of the multiverse led me into naively discussing our spiritual heritage, I will focus on the widely held belief that humans should be seen as primarily spirit beings. In this connection, Alfred Wallace, Jesuit Teilhard de Chardin, B. Haisch and F. A. Wolf were mentioned earlier as being focused on our spiritual nature as well as being able to function studying physical world phenomena. It is worth referring to another spiritually oriented philosopher and sometime scientist, the Spaniard Unamuno, who tended to openly contradict himself on some matters by not rereading what he had written earlier. (This may have been Whitehead's problem).

Unamuno's most recognized book[24] contains some direct and forceful words that are aligned with what others had written about our spirituality. I found a slender statement indicating that he seriously considered the soul to be an essential part of us. He rather harshly wrote,

> I am thinking of those who assign a ridiculously excessive value to life: they do not really believe in the spirit, that is, in their own personal immortality, and so they inveigh against war and the death penalty [precisely because] they do not really believe in the spirit[25].

Others with advanced scientific credentials (such as a working knowledge of relativity physics) switch to a religious environment as a means of balancing their world outlook. In the preface to his book[26], John Polkinghorne seemed to echo the sentiment of Bernard Haisch by pointing out that "the exaggeration of science" at the expense of "other sources of knowledge" ought to be brought into balance. Indeed, I referred to other sources of knowledge in the task of interpreting what is to be found by interrogating the STZM model. There is, however, another issue that would be better handled in a separate book. It pivots on the fundamentally different emphases given to the soul/spirit duality in spiritualistic oriented group practices compared with traditional Christian religious practices.

According to Stewart Goetz and Charles Taliaferro[27], as of 2011, the intellectual climate was still quite hostile toward the idea that we each have a soul. They, however, believe that there is a positive future for it. The authors significantly said, "At some point, an enquiry into the existence and nature of the soul must take up the deep question about the ultimate nature of the cosmos." This is greatly relevant to what I have written in later chapters of this book except that this inquiry interestingly comes from addressing the issue in the reverse direction to mine.

The soul and incarnation clearly go hand in hand, but if you want to focus just on the soul, you can get all the data you need. In *The Republic*, Plato recounted the story of a warrior (probably in the Peloponnesian war) who was pronounced dead from his wounds but who recovered to tell the story of his near-death experience (NDE) .[28] That story must have been about his soul being outside his body. Today, about twenty-four centuries later, the International Association for Near Death Studies[29] is making a video presentation about and for current war veterans who have suffered NDEs while on the battlefield. Most of them were traumatized by the experience and often kept it to themselves for years; even later, they were not able to obtain the appropriate and urgent psychiatric help they needed.

I now focus on other persons of interest who use different approaches to interpreting the world of increasing knowledge that is approaching the condition of overwhelming. Freeman Dyson, the dean of critical logical thinking in many scientific matters, has been associated with a quote that runs something like this: The more and more I look at the world around me the more it looks as if the universe knew we were coming. I have already indicated that the multiverse appears to have been made for us; the proof of this is that plentiful evidence indicates that the spirits (souls) and physical bodies are using it to experience sensations in a place where physical substances are either solid, liquid, or gaseous. Otherwise, everything might seem to be pointless as Steven Weinberg and Susan Blackmore have claimed.

Consider socio-biologist Edward O. Wilson, who wrote the acclaimed book *On Human Nature*.[30] His first "dilemma" (which should more likely be called a conundrum) is that concerning the species' ultimate purpose, he writes, "In a word … we have no particular place to go … The species lacks any goal external to its own biological nature." Here we see again a person confining his thinking to only the physical body rather than the dual body as did Unamuno, Teilhard, Pauli, Wolf, Haisch, Polkinghorne… and many more broad-thinking individuals.

Here is another view. Johann Wolfgang von Goethe believed, "Man is not born to solve the problems of the universe, but to find out where the problem begins and then to restrain himself within the limits of the comprehensible[31]." Perhaps you can see through the portal and recognize that the STZM fulfills the quest—and indeed within the limits of the comprehensible. That is there is no explicit modern physics involved with unintuitive overtones. Acclaimed quantum physicist Werner Heisenberg wrote, "Fact is that which can be described in the language of common sense."[32] I believe I have attempted to adhere closely to this statement in this book. Yet in case you think I am cherry-picking, I have always wondered why quantum physics is not couched in the language of common sense.

Douglas Dean rated precognition's cause as a hard problem. If it is the result of one of the "flaws of nature" John Barrow laid focus on then that gives us a clue in searching for ultimate truth just as some of the laws of nature do at a lesser scale. If the STZM model truthfully shows that there are universes holding our future, we would hardly look for a flaw in the operations of the Cosmos. Rather, the problem might be localized in the sensing and processing system of humans— the mind or subconscious. Once during a consultation with a medical psychic for advice on whether to have a routine (TURP) prostate operation done, I was told she was epileptic. That was the answer to a question that I had *thought to ask*: "How are you doing this?" You can see the dots accumulating. Next questions: is epilepsy a flaw and is there a link between psychic behavior and epilepsy? It turns out that there is a link, and it appears that we have just identified a flaw.

Numerous postings on the Internet indicate that people who

experience seizures routinely experience precognition, premonition, and déjà vu and that the activity of these occurrences is heightened around the times of these seizures. Researchers study the connection between temporal lobe epilepsy (TLE) and psychic behavior, and there is an epilepsy foundation documenting cases on a public blog site. I think it's possible that some psychic individuals may not necessarily be linked to TLE but have psychic ability by some other means.

Early wisdom delegated psychic behavior to a divine source (suggesting it was a virtue). In contrast, the TLE connection is more convincing. TLE has several causes, but many are associated with local brain abnormalities that include cases induced by known injuries in one or more of the temporal lobe areas. There are also documented cases of non-epileptic seizures including the electrical surges in the circuitry of the brain.

It is remarkable that the STZM model has led us to explaining so many long-standing puzzles occurring at different scales. These have tended to appear in the last two chapters of this book because new ideas kept arriving unexpectedly as I wrote. There is one revelation in this book that has special meaning for me though it is not established fact according to the strictest of scientific protocol. The fact is this: I (the one in this universe) am living a life that is essentially copied. A quick but incomplete proof of this is that the future can be seen, and eventually it will with some possible modifications become our present. Therefore, all the events in my life, the crowning one being the rediscovery of the STZM, have already been done before billions of times. The only universe in which one can claim originality of achievements is the Template.

The majority of peaceful-minded people should consider these things.

a. The multiverse and hence our universe possesses a very definite purpose as many people have already made clear to us.

b. Our physical bodies are here expressly for the benefit of spirit energy bodies seeking to experience a physical life.

c. We should keep some timeless words tucked in the back of our minds: "That which is here ... hath already been."[33]

An Extended Bottom Line

This is an eleventh-hour attempt to insert some of the concepts that are still occurring to me as I attempt to end this last chapter. Study the words of Richard Feynman, who encapsulated thoughts of two other famous physicists and a spiritual leader: "What we need is imagination" and "We have to find a new view of the world." Underlying the first quote is a constant theme stressed by Einstein. The second quote is common to statements made by David Bohm[34] and the Dalai Lama. But now, I must have the last word on this matter.

In combining cosmology with undeniable and long-known spiritual information about ourselves, I maintain that I stretched not so much my imagination as my mind and thus my consciousness to the limit, and I became aware of a new view of the world. These are also the words of Flinders Petrie and mimicked by John Dunne. Not only that, the result is a unification of cosmos and spirit. This was achieved at the moment of possession of the Blueprint, which amazingly did not need a single equation for its derivation; it transcends equations. It also turns out that my approach to the initial problem, which follows a minimalist pathway, is quite close to the first mode (of four modes) of Buddhist analytical methodology, which involves "setting up a situation, investigating it, asking one question after another, collecting evidence, and then coming up with a coherent explanation."[35] That sounds like a commonsense approach just as Werner Heisenberg was advocating.

The Geometrical and Metaphysical Content
of *Time and the Multiverse*

Modern physics shares the same domain as preternatural and spiritual phenomena, but the preternatural seems to be based on principles totally different from those of neo-Newtonian physics or relativity physics. I am not sure about quantum mechanics. Pauli could clearly see this. He well understood the existence of another side to life (and that included his life in physics) because he personally experienced a preternatural phenomenon neither he nor Jung could explain. It was called the Pauli effect, which operated like a short, two-element, Jungian synchronicity except that instead of being passive, it was characterized by sudden violent and targeted physical breakages or breakdowns of equipment in labs as well as other buildings whenever Pauli was in the immediate vicinity. Pauli was called a walking poltergeist by his colleagues only some of whom thought it amusing especially when it did not affect them. This was labeled a mind-matter linkage, and it was not confined to Pauli. It is, of course, a very local effect that quickly dissipates.

Jung had his own version of this mind-matter link. As with Pauli, it was associated with breakages typically accompanied by loud noises that were obviously meant to attract attention. Two of these loud events without visible damage occurred in Freud's library in Vienna when Jung was paying him a visit in 1909. They had been talking about precognition and parapsychology in general in which Freud didn't believe. In fact, he was negative toward the occult altogether. At the first of the bangs coming from his bookcase, Freud was stunned. At the second one, which Jung had quickly predicted, Freud was horrified. I call this stunt of Jung's the Jung effect because Jung by his own account knew he was directly associated with the phenomenon.

In quantum physics experiments using the two-slit apparatus, there is the famous reproducible result that human observers can alter the course of the experiment whereas specially designed instrumental detectors cannot seem to do this. In contrast to these weird occurrences, it is a relief to be able to use geometrical constructions available for those who prefer a graphics-oriented approach to presenting a view of our existence. Sir Isaac Newton largely led the way in this area, and modern mathematical physicist Sir Roger Penrose has continued this tradition by using geometrical sketches to support his algebraic constructs. It comes as some relief to be able to join up with part of this tradition in *Time and the Multiverse*.

According to Fleuriot and Paulson,[36] Newton's *Principia* was "a mixture of geometric and algebraic arguments together with Newton's own proof techniques." They said that Newton's reasoning process "displays the impressive deductive power of geometry." Niccolo Guicciardini[37] also commented on Newton's geometric style by saying that the original *Principia* "was burdened by geometrical diagrams [and] almost devoid of symbolical expressions." He offered several acceptable reasons for this; one was related to tailoring *Principia* based on the expected readership.

In my case, I had no option. The prominence of geometric diagrams used to support the text was forced on me by the nature of the material I had to work with. On one hand, I couldn't generate equations as Newton could, and on the other hand, I couldn't afford to dumb down the presentation because it was naturally already approaching that level in many places. Howard Stein[38] wrote about Newton's metaphysics, which will add to my defense against critics.

Finally, we end on the exaltation of geometry given by Pauli in the book he and Jung wrote and published in 1955. He wrote that it was an axiom of the Pythagoreans that "Geometry is the archetype of the beauty of the world."

Notes

1   The statistics are biased on this subject due to it being without a mechanism. A reduction in potential data is caused by people not recalling they dreamt a precognitive dream. The dream memory quickly fades in the first ten seconds if it is not quickly written down as was recommended by John Dunne.

2   M. Hanlon, "World next door," *Aeon Magazine* Nov 6, 2012. According to
     Wikipedia (http://en.wikipedia.org/wiki/Multiverse), February 14, 2015,
     the count, quoting cosmologist Brian Greene, is essentially still the same. In
     his article, Hanlon summed up by writing, "For half a millennium, science
     has been chipping away at the idea that humanity is central and unique. The
     multiverse replaces the chisel with a wrecking ball." This is diametrically
     opposite to what I have concluded in this book. For the serial time-zoned
     multiverse in which each of our lifespans are laid out to be experienced
     within a vast number of universes, it can only be concluded that the mul-
     tiverse was built for souls to experience a physical life. It is an anthropic
     multiverse. Sic fiat.

3   M.J.W. Hall, D. A. Deckert, and H. M. Wiseman, "Quantum phenomena
     modeled by interactions between many classical worlds," in *Physical Review
     X* 4.041013, 2014: 1–17.

4   https://phys.org/news/2010-07-theory-gravity-doesnt-big.html. The above
     report is based on a research paper by M. Banados and P. G. Ferreira,
     "Eddington's Theory of Gravity and Its Progeny," *Physical Review
     Letters* 105, 011101, 2010. A further contribution to the need to replace
     the big bang theory is provided by R. D. Pearson, "Revising Big Bang
     cosmology," in *Speculations in Science and Technology* 21 (1998), 269–76;
     doi:10.1023/A:1005573605781. All these publications support the model I
     adopted for the STZM cosmological model developed in chapter 2.

5   M. Chown, *The universe next door: The making of tomorrow's science* (UK:
     Oxford University Press, 2002). The quote from Richard Feynman comes
     from the foreword of Chown's book, page xi.

6   For example, www.sci-news.com/physics/science-many-interacting-worlds
     and RT.com.

7   Ibid.

8   Ibid.

9   Ibid.

10   Ibid.

11   Ibid.

12   Descriptions and even diagrams offered to describe neo-Newtonian me-
     chanics are difficult to follow indicating that authors were struggling to
     convey their understanding of this strain of Newtonian mechanics, e.g., L.
     Sklar, *Space, time, and spacetime* (University of California Press: *Stanford
     Encyclopaedia of Philosophy: Absolute and relational theories of space and
     motion*, 1974). This was first published in 2006 and underwent a substantive
     revision in 2015. In chapter 2, I did not know of this reference frame to work
     in at the outset. It was only after I had studied the Blueprint that I realized

there was something in there beyond plain Newtonian space-time. Some additional help identifying features of neo-Newtonian space-time came from: A Ashketar and V. Petkov, *Springer Handbook of Spacetime* (New York: Springer, 2014). In what follows, I use space time, which has been my convention for Newtonian type domains. Neo-Newtonian environments are ones in which there is a space domain over and above the space-time region in which events and motions in them are being analyzed; this applies inside universes. Type-1 time originates outside universes and is relative rather than absolute. It also has a manifestation as the Now moment inside universes. Inside universes, there are no absolute velocities, but accelerations (being scalars) can be considered absolute or relative. The movements of universes may be viewed as being relative. It is possible to apply an absolute status, but this is irrelevant here. As in straight Newtonian physics, geometry is Euclidean and it incorporates absolute simultaneity. This is seen as a requirement for the dynamic events line in the Blueprint.

13   Statement made by Lee Smolin, taken from a BBC interview conducted by Quentin Cooper in May 2013. The title of the article was "The mind-bending mysteries of multiple universes."

14   This term was loosely used by early theoretical cosmologists such as Willem de Sitter and George Gamow before the middle of the twentieth century. John A. Wheeler of Princeton University continued its use in the 1960s. The actual term *big bounce*, however, did not enter the scientific literature until 1987 according to the following websites as of December 2014. http://web.uvic.ca/~jtwong/newtheories.htm and http://en.wikipedia.org/wiki/Big_Bounce.

The graph that shows the expansion and contraction of a closed universe can be found in most books on general relativity or on the Internet, e.g., http://www.maplesoft.com/applications/view.aspx?SID=142459&view=html.

15   A mathematically based graph showing the shape of a cycloid is shown in R. Penrose, *The Large, the Small, and the human mind* (Cambridge, UK: Cambridge University Press, 1997), figure 1.16(d) on page 27. In practice, this graph does not fit the facts.

16   M. Bojowald, "Big Bang or Big Bounce? (A) New theory on the Universe's Birth," in *Scientific American*, October 2008, https://www.scientificamerican.com/author/martin-bojowald/.

However, drawing on the simple analogy of a golf ball rebounding off concrete, it is evident that something is needed to cause the initial takeoff. This could simply be another big bang. But even more problematical is what happens to cause a bounce. Then you enter into an endless regression of bangs and bounces. This idea is neither realistic nor testable.

17   See Good, I. J., ed., *The scientist speculates: An anthology of partly-baked ideas* (New York: Capricorn Books, 1965). Good (an Anglicized version of Gudak) was instrumental in cracking the German naval code used during World War II in late May 1941. This action resulted in the sinking of the battleship *Bismarck*. The speculative idea of using a worm tube occurred to me only while writing a third version of this chapter.

18   http://www.dailymail.co.uk/home/moslive/article-1269288/STEPHEN-HAWKING-How-to -build-time-machine.html.

19   Dunne's life could be viewed as a quest for understanding the *go* of things: how to put together by trial and error a flying machine that will be aerodynamically efficient and of course how to get a handle on the mysterious precognitive dream experiences he had that seemed to involve some unusual and perplexing property of time. He was on Priestley's list of "time-haunted" men, and Priestly belongs there too.

20   D. Radin, *Entangled Minds*: *Extrasensory Experiences in a Quantum Reality* (New York: Paraview Pocket Books, 2006).

21   D. Radin, *The Conscious Universe* (San Francisco: Harper Edge, 1997). In addition to being presented with this provocative statement concerning an aspect of precognition, we are given the typical insider exchanges between the people who are perennially skeptical of paranormal (better called preternatural) phenomena and those who perform sincere and carefully structured experiments in the laboratory under controlled conditions.

22   J. D. Barrow, *Living in a Simulated Universe: Universe or Multiverse?* Bernard Carr, ed. (Cambridge: Cambridge University Press, 2009), 481–86.

23   M. Kaku, *Physics of the impossible: A scientific exploration into the world of phasers, force fields, teleportation and time travel* (New York: Doubleday, 2008). My view of his chapter 13 and of calling things impossible is that it is a way out of facing a big challenge. Stephen Weinberg lectured students to go for troubled waters when selecting a thesis topic. Without doubt, our present field of physics is incomplete. Einstein called quantum mechanics an incomplete theory, while his own theory needed help from another recent field, particle physics, which is at the lower end of the size scale where it exists free of relativity.

      One observation in Professor Kaku's section on precognition reveals that he was using improper evidence for the so-called violation of cause and effect, which the multiverse model I described exposes as an illusion. Contrary to this assertion, it was shown that precognitive dreams and visions provide evidence *for* cause and effect to be followed. Does this show that logic and geometry can on occasion efficiently circumvent mathematical physics?

24 M. de Unamuno, *The tragic sense of life in men and nations* (English translation from Spanish, 1912; Princeton, NJ: Princeton University Press, 1972), 291–92, English translation from Spanish, 1912. Unamuno used the terms *soul* and *spirit* interchangeably, but the least confusing usage should be to use *soul* only when the *spirit* is coupled with the physical body.

25 Ibid.

26 As told by J. Polkinghorne, *Beyond Science: the wider human context* (Cambridge: Cambridge University Press 1996).

27 S. Goetz, and C. Taliaferro, *A brief history of the soul* (UK: Wiley-Blackwell, 2011).

28 As told by Dr. Raymond Moody in a UTube video interview. Moody, a physician and psychiatrist, wrote the book *Life after Life*. There, he documented numerous anecdotal accounts of people experiencing out-of-body experiences (OBEs) of which near-death experiences (NDEs) form a class. Thirteen million copies of that book have been sold, and it was translated into a dozen languages. I have experienced an OBE/NDE when I was very young (but not since); a fellow parachute jumper in 1979 had an OBE while exiting the plane on his first jump. It turned out that he did two jumps in one! Wide eyed, he told the whole class when gathered at a debriefing what the experience was like: it was a classic OBE. He—that is his soul—literally "jumped out of his skin." Most dictionaries will not come clean with a proper meaning of this common phrase, but I am providing one here. Again, we see that we are in the same conundrum as we were with precognition; there are never any instruments that are on hand to record these phenomena (if indeed such instruments could be made); nobody else can see what the experiencer sees, and the phenomenon arrives and dissipates so suddenly. The evidence cannot be verified or falsified.

For the book *Life after Life,* Wikipedia quotes, under the section headed Reception, only negative reviews. I think it fair to say, "If you don't play-a-the game, you don't make-a-the calls." But what if you are confronted with the masses of data on these phenomena? What if they have a certain structure? And what if there is a fair chance of an explanation for them?

29 http://iands.org/conference-news.html. The IANDS organization is active in hosting conferences and encouraging new membership.

30 E. O. Wilson, *On human nature* (New York: Bantam Books, 1982), 3. In contrast to this statement, there is a Wilson Wiki quote: "Nature holds the key to our aesthetic, intellectual, cognitive and spiritual satisfaction." Wilson may be recanting here, but what is his definition of spiritual?

31 Goethe (1749–1832) was a German polymath who can be accessed through the Internet.

32   W. Heisenberg, *Physics &Philosophy* (New York: Harper-Collins, 1958), 154.

33   Kohelet 3(15). This is almost correct because if you were in the universe where generic Dunne first thought of his book, then not everything "which is here" will already have "been." This is just a transient technicality. Only a relatively small number of universes are qualified to claim this energetic mini-burst of transient free will. This situation also applies to me because I started this book based on the existence of precognition though I personally experienced it much later.

34   D. Bohm, *Wholeness and the Implicate Order* (London: ARK Paperbacks, 1983). On page 211, he stated, "The fundamental law, then, is that of the immense multidimensional ground; and the projections from this ground determine whatever time orders there may be." This at a guess and along with other statements he makes seems to be pointing through the fog toward a timed-zoned multiverse but alas, the word *multiverse* cannot be found in the index. Indeed, some people called Bohm "infuriating" at times because of his abstruse habit of philosophizing. Could I be accused of that?

35   Dalai Lama XIV, A. Zajonc, et al., *The new Physics and Cosmology: Dialogues with the Dalai Lama* (New York: Oxford University Press, 2004). On pages 34–35, the Dalai Lama stressed that the alternative to the current perception of the world involves finding "a radically new theory ... a totally different view of the world ... It may be time now to bring many other kinds of vision– artistic, religious and spiritual—to bear on these very important questions."

36   J. D. Fleuriot and L. C. Paulson, *Proving Newton's Propositio Kepleriana using geometry and non-standard analysis in Isabelle,* in *Automated Deduction in Geometry* 1998, v. 1669 (lecture notes in artificial intelligence 1999), 47–66.

37   N. Guicciardini, *Analysis and synthesis in Newton's mathematical work,* I. B. Cohen and G. E. Smith, eds. *The Cambridge Companion to Newton* (Cambridge: Cambridge University Press, 2002), 308–28.

38   H. Stein, *Newton's metaphysics,* I. B. Cohen and G. E. Smith, eds. *The Cambridge Companion to Newton* (Cambridge: Cambridge University Press, 2002), 256–307.

APPENDIX 1

Intelligent Design of the Multiverse

Previous attempts to describe a domain or space that might have existed before the big bang or the big bounce have in the last few decades resulted only in very different and generally mutually incompatible scenarios; all of these proposals are speculative without showing signs of being verified. A multiverse could have been planned out in a domain outside the multiverse, but is this the only possibility? Our experience with (intelligent) spiritual entities is that they mingle with us in the same space we use. Those issues aside, the big question is, What is the overall purpose of the multiverse? An answer was put forward in the last two chapters of this book.

We can next ask, is this particular multiverse intelligently designed? This question can be answered using material in chapters 2, 3, 5, and 6 of this book: this appendix represents a summary of that material. Recognition of intelligence is underscored by the well-used dictum that the whole is greater than the sum of its parts.

My role in this undertaking appears rather curious because my soul deliberately incarnated into this body whose life plan included figuring out the geometry of the multiverse[1] and what the reason for it was. In this domain, it projects as a new discovery, but that is an illusion; the discovery had been made many times before going right back to a universe (in practical terms spread over a great number of them—as an event) behind the Template when precognitive dreams were being first experienced. The intelligence, with whom I cannot communicate or even identify, is not concerned with this, that, or any other personal discovery taking place in a copied universe. Yet here I (and all the others who are sharing this life) sit attempting to judge the accomplishments of this intelligence. Following, for the quadrillionth time, is a list of properties (or parts) of the multiverse that strongly suggest it was intelligently designed.

The looped nature of this multiverse allows for recycling of material after it has been in turn spent at the end of a cycle and is then available for regeneration. It thus represents the largest recycling scheme known to us. This requires intelligent thinking and could hardly occur by chance. That might occur in the MWI multiverse, but it now looks highly unlikely.

The Template was a necessary element in the scheme because there obviously had to be a first universe. Organic entities there automatically have a built-in free-will condition with restrictions determined by communities of a particular life form and how well or how poorly they interact among themselves and with other life forms as well as with the environment.

The built-in versatility of genetic material and in particular the ability of RNA to morph into a huge variety of shapes renders these molecules crucial to the process of organic evolution. This process taking place in the Template follows CAPI[2].

The next observation that is of paramount importance is that the once-baffling experience of precognition demands the condition that the universes following the Template must be copies of it—where certain caveats may apply in the very special human realm. This leads to the simultaneous existence of billions upon billions[3] of copies of the same life spread through a section of the multiverse. Every life is preserved as long as the multiverse is kept going. This provides the community of spirits with an extremely large population of physical bodies into which they can incarnate. One group can experience the same physical life simultaneously in many universes that are displaced in time. This process is evidently repeatable by all spirits, hence the term *reincarnation*.

Because of this, the following expressions are equally valid: same soul, many bodies—same body, many souls. This would have particular significance to the community of spirits, but it also projects to us a quality of invariance. Is that an example of ID?

What may be perplexing enough for many people already is the fact that humans are for practical purposes positioned midway between the very small and the very large. Why is this? Take the size range from a neutrino (10^{-24} m) to say the middle of the Virgo galactic cluster (10^{23} m) and consider it to be a deterrent for human interference, misuse, and damage to nature, the history of which already spans a remarkable range of 10^{-10} m (atomic scale) to 10^5 m (depth of oceans to into the stratosphere) where humans exist at a scale $10^{0.4}$ m. The worst infraction so far involves nuclear processes and the danger is in radioactive injection into the biosphere and into the lower stratosphere.

There are some parallel ideas seen here to an idea due to John Gribbin in his book *In Search of the Multiverse*[4]: "The universe is comprehensible to the human mind because it was designed ... by intelligent beings with minds similar to our own ..." Theoretical physicist Paul Davies wrote, "The impression of design in the universe is overwhelming." This implies that the universe has a purpose. In contrast to these verdicts, others[5, 6] have simply said that the universe is pointless, which is to miss the point.

In the early 1980s, particle physicist Alan Guth came up with the cosmic inflation hypothesis applied to the whole universe when it was only 10^{-36} second old. The inflation phase, which was said to be preceded by a super cooling, lasted until the universe was only 10^{-33} seconds old. A number of physicists are not sure if this hypothesis was intelligently designed but we can be sure that whatever did happen was intelligently designed.

This exponential burst of activity, 0.999×10^{-33} seconds in duration, was claimed to be needed to preserve a uniform distribution of particles throughout the universe. The theory is opposed by Sir Roger Penrose and others. In any case, the picture is apparently supported by an elegant mathematical description all packed into an infinitesimally small amount of time. Clearly, in this scenario, the value for Δt would have to be infinitesimal to capture the processes involved. But as I mentioned, some think that a slow start is a more likely scenario than a big bang, which is what cosmic inflation seems to represent. A lower-limit value for Δt would presumably be set by processes in particle physics. That would easily escape human detection (where instruments are excluded). A hunt for activity in brain wave traces typically ends at about 4 megahertz (corresponding to a value of $\Delta t = 0.25 *10^{-6}$ s or about one quarter of a millionth of a second). If no unaccounted for spike in the frequency power spectrum is found in this range, Δt is evidently smaller than the value just given. Other methods such as tracking the growth of crystals from solution would need to be investigated. This may be a naïve approach, but it underscores a venerable human quality: imagination.

Both the cosmic inflation hypothesis and the slow-start hypothesis were the products of human imagination. According to Einstein, "Imagination is everything; it is the preview of life's coming attractions." He also said that imagination trumps knowledge. Actually, both are needed.

Do you see an element of imagination contained in the STZM? While I was at the drafting table producing figure 2.1B, was my mind in resonance with ID? Recall the words of John Gribbin: the universe "was designed ... by intelligent beings with minds similar to our own."

Edgar Mitchell,[7] my next significant character, was an aerospace engineer and an explorer: a dedicated paranormal investigator. I was particularly inspired by his description of being profoundly affected by just the sight of Earth from the Apollo 14 lunar module just as Apollo 8's Bill Anders and Frank Borman were affected when they both saw and famously photographed earthrise while the module was orbiting the moon. There is an important link here to the realization of an almost limitless breadth, depth, and power of human consciousness. Without taking a single measurement, it is possible for us to acquire an instant appreciation and significance of our very existence on a dot-sized planet in a huge space. Edgar wrote how he suddenly knew very much more about the significance of his existence and particularly about communication between people separated by immense distances.

My own (low-budget) experience along this line of thinking occurred when I had just finished figure 2.1B. Stepping back from the drafting board, I suddenly realized I was looking at something that transcended calculations. In my mind, I had left the now totally invisible planet, our solar system, and our galaxy; our own universe was a mere average sphere among many. I was looking down on the entire multiverse. So again, we're back to the question of what is behind this magnificent scheme.

Conceive of asking the right questions about the nature of our universe, then include the multiverse and seriously think about the possibility that this incredibly coherent Cosmic structure was constructed solely for the benefit of our souls not to mention also for the benefit of other organic life containing souls that may lurk somewhere in the cosmic space-scape. I will end by saying that Bernard Haisch[8] has inadvertently supplied me with the idea for another of the many possible titles appropriate to this book: *The Purpose-Guided Multiverse*.

Notes

1   The MIW multiverse (chapter 6 note 6) possesses a number of connections to the STZM such as: each has a large but finite number of universes, and each has the same physics in its universes.

2   Compute and paste in.

3   As with like expressions, this should be taken just as an expression for a very large, unknown number.

4   J. Gribbin, *In search of the multiverse* (London: Penguin Books, 2010).

5   This refers to prominent mathematical physicist, Steven Weinberg, the author of the popular book *The first three minutes* (1977). The book had three printings in 1977 alone, indicating the interest in such a seemingly far-out topic describing what happened for three minutes near the start of our universe's theoretical cosmic clock 13.7 billion years ago. He also

remarked, "The more the universe seems comprehensible, the more it also seems pointless"; https://www.goodreads.com/author/quotes/86758. Steven Weinberg. Here is more of this negativity: "The effort to understand the universe is one of the very few things that lifts human life a little above the level of farce, and gives it some of the grace of tragedy." Contrast this view with Freeman Dyson's view, "The more I examine the universe and the details of its architecture, the more evidence I find that the universe in some sense must have known we were coming."

6   I quote from http://www.thirdwaymagazine.co.uk/editions/november-10/high-profile/do-we-copy.aspx, an interview with psychiatrist Susan Blackmore.

Peter Moore (interviewer): "You have said several times that we have no purpose …" Blackmore: "It seems to me that, as far as I can tell, the universe has no ultimate purpose—it's pointless. We are here just because it so happens the laws of physics are the way they are, and evolution is inevitable given the way it is."

7   E. D. Mitchell, *Psychic Exploration: A challenge for science,* introduction; J. White, ed. (Florida: Capricorn Books, 1976).

8   B. Haisch, *The Purpose-Guided Universe: Believing in Einstein, Darwin, and God* (New Jersey: New Page Books, 2010).

Reflections on Precognition Research a Century Ago

W. G. Roll[1] noted that Dr. Eugene Osty[2] quoted Charles Richet[3] as saying that precognition was "paradoxical, strange, absurd and real." Major publications containing studies on precognition in addition to many other paranormal phenomena had appeared in English and French while Dunne was a soldier, then an aeronautical engineer and aviator, and during his search for a geometrical explanation of time that culminated with his book *An Experiment with Time* in 1927. He appeared totally unaware of the large amount of paranormal research activity being done on either side of the English Channel and elsewhere. French publications were translated into English several years before Dunne had apparently handed in the final version of his manuscript to Faber and Faber in London. The book did not mention any of these activities. To my knowledge, only R. L. Mégroz (see chapter 4) made an issue of this.

Despite this, Dunne's book attracted an unduly large amount of attention in England because to some it looked as if he had discovered a new phenomenon (precognition) and had explained the nature of time shifts mathematically. However, there were other avid readers among the then-aging Victorian-era population who were either unaware of this phenomenon or who had been suppressed from discussing the subject even though they might have experienced only fragmentary dreams that seemed to carry a message of some kind. Mégroz tried to tell readers of the *Times Literary Supplement* that he intended to have published an anthology of dreams that would include precognitions compiled from many modern and historical sources. It was published as *The Dream World* in 1939.

Nobody has seriously challenged Dunne's dream accounts, which do contain valuable insights into the nature of precognitive dreams though only some of them are found to be genuinely precognitive. They formed the basis for Dunne's assault on the problem of how to explain this phenomenon. He immediately saw that precognitions involved images displaced in clock time from the present clocks, and this triggered his attempt to model it. This also occurred to me independently eight decades later. In my case, however, I hadn't experienced obvious precognitive dreams, but then, I didn't keep a dream log when I started the multiverse project.

My first dream log entry was on November 4, 2011, using early twentieth-century protocols to capture my first precognitive dream (of the Bow River flooding in Calgary) on the morning of April 10, 2013. It was partly symbolic, had no warning value, was quickly forgotten, and discovered only much later when I was checking out another logged dream (number 10, July 4–5, 2013) that I had remembered. The actualized event occurred between late June and early July 2013. It was after this that I realized the great value of having a dream log and of having my own experience of the phenomenon of precognition. That simple protocol represents a legacy of John Dunne.

Notes

1   W. Roll, "Transcendental Mind: Eugene Osty's Supernormal Faculties in Man," *J. Scientific Exploration* 19(4) (2005) 615–24.
2   E. Osty, *Supernormal Faculties in Man: An experimental study*, translated from the French by Stanley De Brath (New York: Dutton, London: Methuen, 1923). Osty was a well-known French medical doctor who devoted a large part of his later career to investigating paranormal phenomena. With the help of his son (a physicist), he built devices using light beams to detect

trickery in psychic test subjects as well as to help determine a mechanism for paranormal phenomena. Osty assumed the directorship of the Institut Métapsychique International in 1924 and was a corresponding member of the Society for Psychical Research, London.

3   C. Richet, *Thirty Years of Psychical Research,* translated from the second French edition (New York: Macmillan, 1923; C. Richet, *Our Sixth Sense* London Rider Transl. from French, 1928; C. Richet, *Traité de métapsychique* Paris, F. Alcan, 1922; J. Maxwell and C. Richet, *Metapsychical phenomena: methods and observations,* London: Duckworth, 1905). In 1905, Richet was named president of the Society for Psychical Research in London. He became a doctor of medicine in 1869 (at age nineteen), doctor of sciences in 1878, and professor of physiology from 1887 on in the faculty of medicine of the University of Paris. In 1913, he was awarded the Nobel Prize in medicine for research on anaphylaxis. He was also a pioneer in early French aviation. In 1914, he became a member of the French Academy of Sciences. In 1919, he became honorary president of the Institut Métapsychique International in Paris, and in 1929, its full-time president.

Two Famous Cases of Premonitions by Lady Kathleen Kennet

Introduction

I have included this appendix to show that there is another side to almost every great story. The well-known books about Scott of the Antarctic and about Mallory of Everest do not include the other, just as important nonphysical side.

Kathleen Kennet was the former honorary Lady Scott, wife of Captain Robert Scott. Kathleen was an established psychic. The cases I know about involve Robert Scott (d. 1912) and George Leigh Mallory (d. 1924). Both perished under tragic circumstances that had been foreseen. This involved three phenomena: precognition, premonition, and telepathy. In the Robert Scott–Kathleen Scott interactions, precognition, premonition, and one-way telepathic connections seem to have been involved. The quite different case of Mallory involved one arranged reading of his near future on the basis of a referral from friends who knew Kathleen.

While I have been previously concentrating on precognitions, which in principle provide more-explicit information than do premonitions, I have had to use the latter here because of the different subject matter and the type of documentation involved. Only the mention of a "bad dream" (about Robert Scott) experienced by Kathleen suggests visual information was occurring. Precognition took on a secondary role to premonition here. It thus seems that precognition and premonition may be two aspects of a phenomenon arriving via a single channel.

Brief Biography of Kathleen

Kathleen was born Edith Agnes Kathleen Bruce in England in 1878, and yes, she said she was related to Robert the Bruce. In her early years, her adventurous spirit was already developing; she also became a respectably well-known sculptor and equally well-known, high-stratum socialite in London. In September 1908, after a faltering start, she married Captain Robert Falcon (Con) Scott. The next year, Peter Markham Scott was born.

Among a large network of friends and acquaintances were such notables as George Bernard Shaw, J. M. Barrie, Herbert Henry Asquith (prime minister of England 1908–1916), Sir Max Beerbohm, Fridtjof Nansen, Guglielmo Marconi, Isadora Duncan, Auguste Rodin, and even aircraft baron Thomas Sopwith (later knighted). She often had complimentary rides in some of the new aircraft designed to enter the air war against Germany.

Kathleen was a classic but classy tomboy. I suspect her psychic talents had something to do with some of her rather unusual social connections as well as connections related to her fundraising for the Antarctic expedition, which continued even after the members of the polar party had met their fate.

Captain Robert Falcon Scott RN

The documentation of this case history is unique because Kathleen and Robert (nicknamed Con) kept diaries that overlapped from November 29, 1910, to March 29, 1912. It was thus a perfect setup for documenting long-distance connections between two people. Afterward, they had mutually agreed to read their diaries to one another. Note that they were separated by the International Date Line. The Antarctic sector in which the expedition was operating is a day ahead of Greenwich, which Kathleen's diary dates apply to.

In 1910, Kathleen traveled to New Zealand via South Africa and Australia to bid farewell to her Antarctic-bound husband, whom she never saw again. Returning to England, the long wait began—two years and seven weeks. The first inkling of premonitions comes from Kathleen and her two-year-old son. She wrote in her diary on September 20,[1] 1911, "Rather a horrid day today. I woke up having had a bad dream about you, and then Peter came very close to me and said emphatically 'Daddy won't come back', as though in answer to my silly thoughts."[2] She wrote more yet in a style that seemed to be for the case that Con was returning—but it was as if she was feigning the entries, as if she had felt early on the expedition would meet a tragic outcome. Add to that—she never addressed her husband by name or endearingly.

The Trek to the South Pole

Scott had already been to the Antarctic as leader of the Royal Geographic/Royal Society Discovery Expedition in 1901–04 during which with Ernest Shackleton and Edward Wilson, he then came within 850 kilometers of the South Pole. He had thereby established himself in England as a polar explorer. He was thus on his second, and last, quest for the pole.

In England on September 20, 1911, equivalent to the 21st where Con was, he was on a 175-mile sledge training journey with Birdy Bowers and feeling quite content with life as Con recorded in his diary.

The premonition (as I will call it) was delivered six months plus a week (within a day or two) before Con died. The following paragraphs will lead up to this event.

Through all of 1911, the expedition's prospects looked favorable; the entire caravan of people, sledges, ponies, and supplies meant to put five men of the British Services at the South Pole departed from the wintering-over hut at Cape Evans on November 1, 1911. By January 4, 1912, the group led by Scott was making the final assault. They reached the hard-won pole five weeks after the Norwegians, who had left a flagged tent there. The bitter return trek as far as it went was surely the worst journey in the world rather than the short but very cold journey to the emperor penguin colony at Cape Crozier in the winter of 1911. That journey was endured in the interests of biological science by Wilson, Bowers, and Cherry-Garrard. It was C-G who wrote the well-known book with just that title.

By the middle of February 1912, the returning polar party's situation was deteriorating; they were running low on food and fuel, and several men were suffering injury and frostbite especially Edgar Evans. His was the first casualty; he died due to an injury sustained in a crevasse accident on February 17. This must have caused Scott and the others very great concern. Even Scott must have been thinking seriously about his own chances of survival at that point. Leaving Evans's body in a snowy grave, they stumbled on over a difficult sastrugi-plastered snow surface. The following day, they dined on pony meat at the next depot.

The Telepathic Connection

On February 18, 1912, Kathleen made an entry in her diary: "Very taken up with you all evening, I wonder if anything special is happening to you. Something odd happened to the clocks between 9 and 10 pm."[3] The next entry was on the 19th: "Still rather taken up by you and a wee bit depressed. As you ought about now to be returning to [the] ship, I see no reason for depression. I wonder."

That day approximately, Con made a comment in his diary, "I wonder what is in store for us, with some alarm at the lateness of the season." This is loosely interpreted as a telepathic communication made without exact times and dates that were within a day plus or minus. Notice that they both used the words "I wonder"; Kathleen used them twice. Note that she was maintaining a diary monologue directed toward Con; the reverse apparently wasn't occurring.

On February 20, Kathleen wrote that she "felt rotten"[4] though she was still quite involved in expedition fundraising that day. Meanwhile, Con wrote, "Pray God we get better traveling as we are not so fit as we were, and the season is advancing apace." The daylight was shrinking, and Con wrote, "Heavy toiling all day" inducing "gloomiest thoughts at times … We never won a march of 8 miles with greater difficulty, but we cannot go on like this."

At that point, it was March. Lawrence Oates was showing signs of severe frostbite and leg infection. On the 16th, they lost Oates, who finally went outside the tent to perish. On March 29, Scott penned his last entry.

Kathleen had evidently not been maintaining the telepathic connection, but it is easy to see why. She had busied herself in many physical activities such as "sculpturing, skating and flying." Her active friendships in a well-established social network included a large percentage of men, the most prominent being Prime Minister Asquith. That took her mind off worrying about Con. She created "distractions, and a cheerful resistance to any show of concern." This would surely cause suppression of any psychic activity. But it seems she had already known for a long time that something tragic had happened. Indeed, ominous signs were appearing.

On March 6, there were conflicting reports reaching London about who had reached the pole first, Amundsen or Scott. On March 8, Peter told his mother, "Amundsen and Daddy both got to the pole. Daddy's stopped working now." In fact, "about the 21st March" (the 20th in London), the last surviving three "were laid up by a blizzard only about 11 miles from One Ton depot." They all died of slow starvation and dehydration in the tent eight days later. So Peter's reading was a clear two-part statement of fact and a premonition with roughly a twelve-day lead. This could have been a very explicit precognition without any symbolism making it easier for a youngster of just two and a half to relay his impressions to his mother. Or there is the possibility that Peter was getting all this information from his mother.

I will now try to analyze a difficult-to-comprehend note from

Kathleen to Con that was found inside Scott's diary by the search party the next spring. It was written "on a torn piece of paper" in pencil; it read,[5]

> Look you—when you are away South I want you to be sure that if there be a risk to take or leave, you will take it, or if there is a danger for you or another man to face, it will be you who faces it, just as much as before you met [Peter] and me. Because man dear we can do without you please know for sure we can … I love you more than I thought could be possible, but I want you to realize that it wouldn't be your physical life that would profit me [and Peter] most. If there's anything you think worth doing at the cost of your life—Do it. We shall only be glad. Do you understand me? … How awful if you don't.

This is a most remarkable and yet the most appallingly insensitive if not clinical communication of its kind. I think it is an expression of knowing about destiny, which as you have seen is the key element of understanding how we are connected to the multiverse. It looks as if Robert Scott had emotionally detached himself from Kathleen and played out his role as a navy man.

It has been established that mother and son were psychically gifted. There is nothing unique in that.[6] What is remarkable is that Kathleen seemed to be fully aware she was also acting through a life drama with another person; both their lives were fully prescribed just as in Shakespeare's *Romeo and Juliet*, where your roles are already defined and the script has already been written. Susan Blackmore, as you read in Chapter 3, confirmed this by claiming that she lived completely accepting that she had no free will.[7]

Kathleen was traveling by ship to New Zealand in mid-February 1913 to rendezvous with the *Terra Nova* at Lyttelton. While still at sea on the 19th, the captain summoned her to his cabin and nervously handed her the wire message that said Captain Scott had died in a blizzard after reaching the South Pole on January 18, 1912. She physically reacted as if she already knew that this had happened and went about her usual activities "to keep her mind off the subject."

Later, she wrote, "Had he died before I had known his gloriousness, or before he had been the father of my son, I might have felt a loss. Now, I have felt none." She even wrote, "Won't anybody understand that? ... Probably nobody." She just went on with her life forbidding any grief to obstruct her. She was so effective in this that George Bernard Shaw "wrote in a letter that she did not seem to feel her loss at all."[8]

George Leigh Mallory

In note number 2, there is no mention of Kathleen meeting with George Leigh Mallory, the British mountain climber who mysteriously disappeared on Mount Everest in 1924. For this, I had to switch to two other books.[9, 10]

There are some strange parallels between Scott and his Antarctic saga and Mallory and his Everest saga as you will see though this is peripheral to the main intent of the appendix. Mallory was a member of the first two British Everest expeditions in 1921 and 1922. The 1921 expedition was a reconnaissance, but the second one was a serious attempt to attain the summit, which they missed by over 500 meters.

In May 1923, Mallory was asked by the organizers of the third British Everest expedition to join it and later to compile a list of candidates for the planned 1924 climbing team; he included his name (as "self") but with three question marks after it.[11] George was still undecided in early 1924 as the departure date loomed. The secretary of the Mount Everest Committee Arthur Hinks kept up some pressure on Mallory to definitely commit himself. But there were some financial, job, and family concerns about George's participation. Besides, George's wife, Ruth, was having unpleasant forebodings, and there was some evidence that even George had serious doubts. He was good friends with Geoffrey Winthrop Young,[12] whose brother, Edward Hilton Young,[13] was later to marry Kathleen Scott.

Mallory went with Geoffrey and his wife, Len, to see Kathleen.

It was a set up for the obviously worried Mallory to get some private consultation from Kathleen. This is an interesting link, but unfortunately, no one has provided a detailed account of the meeting itself, so the following secondhand accounts are all we have for building a case for predestination. All the information comes from statements made later by George Mallory to his closest friends.

It was reported that after the meeting, George had said, "that he did not want to return to Everest again." Close friend Geoffrey Keynes, whom George confided in over the same subject, had reported him saying much the same thing. Moreover, he said to Keynes that "what he would have to face would be more like war than adventure, and that he did not believe he would return alive." He "knew that no one would criticize him if he refused to go, but he felt it a compulsion."[14] He argued that he owed a duty to the expedition. This eventually overrode the duty to his wife and children.

Geoffrey Young and George's sister, Mary, urged George not to go on the expedition.[15] Ruth's forebodings increased as the departure time approached. Yet she did not insist that George not go despite George having told her, "You must tell me if you can't bear the idea of my going again and that will settle it anyway."

To make matters more complicated, Mallory had recently just been offered a traveling lectureship position based out of Cambridge; moreover, the job had been arranged by none other than Arthur Hinks. George had accepted the job offer and was suddenly in the delicate position of having to ask the Rev. David Cranage, his new, immediate superior, that he wanted to take leave before even starting the job. After George told Hinks that he was having a horrible time on a tightrope, Hinks stepped in and engineered a solution to the dilemma. George was offered six months' leave at half pay. Evidently, that and Ruth's lack of sufficient resistance sealed George's fate. His story is arguably as suspenseful and as famous as that of Scott's Antarctic saga of just over a decade earlier.

One thing stands out here in both these stories. It is the interconnectedness of people surrounding one prominent person. The networking just keeps on expanding in fractal fashion; though the original context fades, the effects keep rippling out. This helps to understand the words of David Bohm when he seemed to reach out for us to believe in the concept of totality, where everything is connected. Otherwise, he said, "We are guided by a self-willed view ... Such a view is false."

Consider the following. The last words written in Scott's diary reached out to the British nation: "For God's sake look after our people." Nothing was personally directed toward Kathleen. But she was already networked with a large and influential group of British people in high places that such imploring words written by a dying British navy captain were hardly necessary. But then, imagine if Robert Scott had defied the predictions and had come face to face with a wife who had written those seemingly heartless words on a torn scrap of paper. To me, this indicates that Kathleen must have been very sure of her prediction at a very early stage.

With Mallory as with Scott, the situation was quite complex: both would have been thinking of knighthood. This was demonstrated immediately upon Edmund Hillary's return from Mount Everest in 1953. There were no strong pressures working against Mallory going to Mount Everest, so he could be looked upon as being prepared to take the risk. Despite predictions, he would have been tricked by the illusion of free will just as W. T. Stead was led to sail on the *Titanic* despite all the indications he should have canceled. He was one who definitely could not break loose of the grip of *la forza del destino*.

However, regardless of what it looks like and even if you strongly believe in free will, it was demonstrated in this book that people in copied universes basically have no free will. (There are occasional and local exceptions to this—the Dunne effect—as I pointed out in chapter 5.) This illusory free will is merely copied free will. True free will is regularly experienced only in the Template. That effect and the results are then permanently incorporated in all subsequent copies of the body and surroundings thus affected. So our Scott and our Mallory (of this universe) were simply acting out the script that was set for their physical bodies.

Notes

1 Because of the intervening International date line, the date where the polar party was would have been September 21.

2 L. Young, *The Life of Kathleen Scott* (London: Macmillan, 1995). Kathleen's and Robert's diary entries were intended to be shared upon his return. That is naturally what she would have said to him though silently believing that might not happen.

3 Ibid.

4 Ibid.

5 This very clinical and insensitive note shows that Kathleen was quite blunt with her message, suggesting that she was confident about her premonitions and one precognition. I happened to come across a revealing interview with an actress, Jenny Coverak, who portrayed Kathleen in a one person play, "A father for my son." The play was based on material in Kathleen's diaries. Coverak made the following statements in answer to the interviewer: "Kathleen is quite an amazing character. She wanted a son but not a husband and so was looking for the right man." George Bernard Shaw told her, "No woman ever born had a narrower escape from being a man. My affection for you is the nearest I ever came to homosexuality ... she was so unorthodox; she was a tom-boy, a free spirit and full of fun, always finding the positive in situations," http://budleighbrewsterunited.blogspot.ca/2011/04/woman-who-knew-her-own-mind-interview.html.

6 Psychic ability often but not always runs in families. I have spoken to three unrelated, fee-charging psychics who are members of a lineage and one non-fee-charging sensitive whose mother was psychic. See http://www.psychicsuniverse.com/articles/spirituality/psychic-abilities/psychics-are-they-born-or-made. Some psychics are epileptic; some become psychic after head injuries.

7 J. Brockman, *What we believe but cannot prove* (New York: Harper Perennial, 2006).

8 L. Young, *The Life of Kathleen Scott* (London: Macmillan, 1995). Kathleen's and Robert's diary entries were intended to be shared upon his return. That is naturally what she would have said to him though silently believing that might not happen.

9 P. and L. Gillman, *The wildest dream* (London: Headline Book, 2001).

10 W. Davis, *Into the silence* (New York: Knopf, 2011). These last two books frequently quote from the same reliable sources that I have not felt the need to access.

11 P. and L. Gillman, *The wildest dream* (London: Headline Book, 2001).

12 G. W. Young was one of the elite British rock climbers in the early part of the twentieth century. He lost a leg in World War I, but with a prosthetic, he famously managed to climb the Matterhorn.

13 E. H. Young was the younger brother of Geoffrey. Hilton, as he was known, lost an arm in World War I. Speaking at a school prize-giving on July 13, 1923, Young "recommended the boys to go in for great risks and dangerous deeds. Let them have adventure, and the madder the adventure, the better," http://en.wikipedia.org/wiki/Hilton_Young,_1st_Baron_Kennet. Hilton was already inside the British political system when he married Kathleen Scott. In 1935, he was made 1st Baron Kennet, making Lady Scott, this time, a baroness—Lady Kennet.

14 W. Davis, *Into the silence* (New York: Knopf, 2011). These last two books frequently quote from the same reliable sources that I have not felt the need to access.

15 Ibid.

TABLE

Key dates in the Robert Scott Antarctic Saga

1910	1911	1912 ↓	1913
↑ The saga begins 29 Nov. 1910 Capt. Scott departs NZ for the Antarctic and Kathleen returns to England.	↑ Premon. Precogn. 20 Sept. GMT. Kathleen and Peter get advanced warning of the tragedy.	SP ↑↑29 Mar. NZST 8 Mar; 29 Mar. Scott † ↑Peter's premonition of the end of the journey and being correct about his father reaching the South Pole (SP)	↑ 19 Feb. Kathleen informed of her husband's death almost a year before.

APPENDIX 4

J. B. Priestley's Analysis of Anecdotal Psychic Experiences

Introduction

John B. Priestley (1894–1984) was a well-known English literary figure in the early to mid-twentieth century. Despite his many different activities, he spent considerable effort trying to understand the nature of time. He recognized some writers as time haunted as mentioned in chapter 4, and he was of that ilk himself but being more like time obsessed. He wove themes involving time into several of his plays, which were inspired by his reading of John Dunne's book *An Experiment with Time* published in 1927.

Later in life, he decided to write a book about man's entanglement with different aspects of time. Instead of using Dunne's already well-known dream records, even his own limited ones, or the many others available by that time, he sought to obtain a fresh collection by use of the latest media available at that time—BBC television.

In 1963, a TV interview was arranged in which Priestley gave viewers a presentation on the research involved in writing his book.[1] This included the apparently anomalous aspects of time. At the end of the interview, on Priestley's behalf, the interviewer (H. Wheldon[2]) asked viewers to send Priestley "any experiences that appeared to challenge the conventional and commonsense idea of time."

Priestley wrote,

> The response was so immediate and so generous ... I spent days and days opening letters ... There was one letter that stood out, and I am going to reproduce it here because I think that there are many classically educated people around today that need to rethink their attitude towards time and what may be called time warps.

The author of the following letter is unknown.

> Dear Mr. Priestly: I wish I could agree with you that the general climate of opinion is now becoming sympathetic towards a serious study of time. When your fascinating conversation with Huw Wheldon was televised recently I was staying with relatives in Wales and whenever the topic was announced the room emptied rapidly ... One of the group, who was "an engineer and technologist" remarked: "For heaven's sake don't sit around listening to that kind of thing. It just makes people morbid—or drives them around the bend ...I wondered ... why ... normal intelligent people take to their heels at the suggestion that time be considered and even investigated. Is it because most of us are already aware of strange aberrations in our own experience and react to mystery with primitive fear? ... Having succeeded in transforming this planet into an animated horror-comic and now turning acquisitive eyes towards other worlds to conquer—haven't we the sense to realize that some of the most intimate and significant fundamentals of our own human nature are still completely unknown, and that it is time something was done about that?"

This was for me as well as Priestley a very good assessment. Notice that this man had an intuition that it was worth studying the nature of time and the nature of our existence together—which is what this book has covered. Recall John Wheeler, the theoretical physicist. He wanted to rid AAAS[3] of the pseudos, the paranormal investigators who had just been admitted into the halls of American science? He failed to achieve his requested rout. Is there still some hidden resistance in America?

In later life, Wheeler seemed to soften his attitude toward these matters as indicated by what he wrote about the human consciousness and how its presence can affect subatomic particles though that idea wasn't original to him.

That letter from an unknown correspondent evidently resonated with Priestley, for he said, "What this letter makes clear—and I could support it with hundreds of others—is the way in which the Fortress[4] exerts its power and uses its influence." He then launched into his own rant on this subject that I will omit because he made a very significant enough next point: "Many of the people who wrote to me confessed that they had been afraid of mentioning to anybody the queer Time experiences they had had—often ... taking the form of precognitive dreams." He mused over how many other people, perhaps millions, simply refused to remember some time anomaly and just "shrugged it away" as being an irrelevant experience. Priestley had an original method of classifying his data.

Letters Were Grouped in Categories

Over a thousand letters were received from correspondents aged eighteen to eighty, and some stories went back half a century. Priestly had the enormous task of sorting out letters into piles according to a classification system he devised. Well aware that there were already enough documents and books about time anomalies in the literature "to fill—and possibly sink—a canal barge," he felt his letter project would be more rewarding. Looking at his results and interpretations, I agree. After eliminating the crank letters, there were five (A to E) categories established.

A. This was for possible examples of what he called "the influence of the future on the present." He excluded cases of dreaming. This was significantly "the smallest category" though he did say that had he explained this type of phenomenon better, he might have received more letters. This category was termed FIP, and it would likely have contained accounts that supported or appeared to support the *Titanic* effect where data were arriving as premonitions or visions.

B. This was for the most convincing precognitive dreams; there was a large collection in this category. However, I assume that even these would not meet the excessive standards of the SPR in London or the ASPR in New York. There is no indication that this overly influenced Priestley just as it did not worry John Dunne. The ratio of women to men responders was about three to one, and the same held for the next category.

C. This was for "precognitive dreams that were not clearly stated and not sufficiently trustworthy, for premonitions and queer hunches that came right, and of odd little Time experiences not easy to explain." They seemed mostly to be sincere accounts; the vast number of these (600–700 letters) was to Priestly "very significant." This category easily held the largest population of letters. It was here that women complained that their husbands were not interested in their dreams and therefore they were eager to tell someone who would listen to them.

D. This was for letters in which people recommended books to Priestley. It was a small category and not important to the main survey. It would surely have included Dunne's book, which Priestly was already familiar with.

E. The letters in this category contained opinions. I understand that they analyzed individual stories or groups of stories and tried to argue for a natural explanation. They were written mostly by men, who were typically academics. "Few were interested in theories" (such as, I assume, the one Dunne put forward).

Some people identified space rather than time as the intractable problem of where to locate these dreams or seen events. That hits the nail on the head as I hope this book has shown. But Priestly didn't seem responsive to this idea; he put his bets on a modified Dunne model with three levels of time, but that was merely an arbitrary truncation of Dunne's rejected infinite serial time, which is invalid regardless of how many terms are put into it. Priestley actually knew Dunne quite well and had him lecture to the cast who acted in Priestley time plays. They would have been based only on Dunne's dreams.

Some of the Stories and Analyses

Dreams often contain special features that are very uncommon, and if there are several of these features and the same number in the waking experience, this strengthens the claim that a precognitive dream has occurred. To show that he was also playing the game, Priestley cited[5] two dreams he had had in which such unusual features played a pivotal role in convincing him that the dreams were precognitive. I will deal only with the second dream, which occurred in about 1925.

> I found myself in the front row of a balcony or gallery in some colossal vague theatre … On what I assumed to be the stage, equally vast and without any definite arch, was a brilliantly colored and fantastic spectacle, quite motionless, quite unlike anything I had ever seen before. It was an unusually impressive dream which haunted me for weeks afterward … In the earlier 1930's I paid my first visit to the Grand Canyon, arriving in the early morning when there was thick mist and nothing to be seen. I sat for some time close to the railing on the South Rim, in front of the hotel … Suddenly … the mist

lifted … and then I saw, as if I were sitting in the front
row of a balcony, that brilliantly colored and fantastic
spectacle, quite motionless, that I had seen in my dream
"theatre" … My dream of years before had shown me a
preview of my first sight of the Grand Canyon.

For the most part, however, Priestley related many very convincing stories from the B category letters picked at random. These are the
primary ones relevant to this book. They are in a sense backed up by
the letters in the C category though they were qualitatively judged less
reliable than the letters in the B category. However, I want to quote
from one file of Priestley's data bank that was not from the mailed
letters. I selected it because it was very relevant to two of my dreams
and one that hasn't been verified yet.

Priestley related that a neighbor and friend of his, a university
lecturer, visited him two or three weeks after he had sorted out most
of his letters. The friend had made notes on some prescient vision experiences of his, but they belonged to the C category. Because Priestley
knew the man to be intelligent and scrupulously truthful, and noting
that he said the experiences were disturbing enough for him to have
hidden them away for a very long time, Priestley relaxed his skepticism. The ten short, concise sketches involved a violent death of a person whose name appeared in print on a label above a scene indicating
the type of death (airplane crash, car accident, etc.). Such vignettes
arrived days to weeks before the actual event occurred as reported in
the media. The friend had no connection whatsoever with any of the
disaster victims and the vignettes seem to have come to him spontaneously as visions while he was engaged in some physical activity.

I had never heard or read about this type of communication in
dreams before, but that was about to change. First, take note of some
dates. I ordered Priestley's book online from ABE books on November,
18, 2009, and I received it on December 9. I recall spending several
months dipping into it especially his misinterpretation of Dunne's
serial-time model. I would also have read the chapter dealing with his
very interesting presentation of dream types by early 2010.

Weird as this may sound, I have had two dreams since then where labeling appeared as a means of communicating information. I had remembered Priestley's story about his neighbor's ten visions in which name labels appeared above a scene depicting a particular type of disaster. The following was taken from my dream logbook. It was the second of my two dreams so far that involved labeling. At 05:20 on November 15, 2011, I wrote out my dream. I was in the company of three people. A French boy (by the feel of it) said to me, "You made a mistake in your ... integration." I looked down as if distracted and saw a label sticking up out of a patch of clover. Printed clearly on the label was the name Dirichlet. I remembered the name from an applied mathematics course I had taken at the University of Canterbury in 1962. Dirichlet was a celebrated French student of the famous Gauss, whose multipurpose bell curve equation I used to form the torus manifold in the STZM. I did some engineering on it to get a synthetic three-dimensional manifold. Unknown to me then, Dirichlet had already achieved a three-dimensional rendition of the Gauss formula analytically resulting in the symmetric Dirichlet distribution, and it involved integrals! However, there is no point in me going back to make changes as even the Dirichlet formulation would still need some engineering done on it.

Priestley's mail may sound like a huge number of letters on such a sensitive and suppressed subject, but an editor of the *New York Times*, Robert Nelson, repeated the procedure of soliciting time-related correspondence from readers of the newspaper. He collected about 5,000 cases of premonitions and precognitions of which he judged fifty as being convincing hits. Half those hits came from only five correspondents.[6]

A Distinguished Witness to a Vision of the Future

I will finish this appendix with two cases involving Air Marshal Sir Victor Goddard. I have found that there are many distinguished and well-known people who have been involved in precognition as well as past cognition experiences. Here the first case involves a prevision; the second involves a dream precognition experienced by another recipient whom Goddard happened to meet.

In 1935, then Wing Commander Victor Goddard[7] of the Royal Air Force experienced a time shift while piloting a Hawker biplane.[8] He was flying solo from Edinburgh, Scotland, to his home base in Andover, England. Deciding to fly over an abandoned airfield at Drem, he found the airfield was overgrown, the hangars were dilapidated, and cows grazed where airplanes had once parked. Continuing his flight toward Andover, Goddard flew into a bizarre storm whose high winds and strange, yellow-brown clouds caused some loss of altitude; his plane began to spiral down. Regaining control, Goddard found that he was headed back toward Drem. As he approached the airfield, the storm suddenly dissipated; he was suddenly flying in clear weather. That time, the airfield looked completely different. The hangars looked new. There were four airplanes on the ground; three were familiar biplanes but painted an unfamiliar yellow. The fourth was a monoplane, which the RAF did not yet have in 1935. The mechanics were dressed in blue coveralls whereas all the then-current RAF mechanics were issued brown coveralls. Additionally strange—none of the mechanics seemed to notice him fly over. Turning around, he again encountered the storm but eventually managed to land at Andover. It wasn't until 1939 that the RAF began to paint their planes yellow, acquired a monoplane of the type Goddard had seen, and issued blue coveralls to mechanics.

This story is (relative to the present) a future counterpart of the story related in *An Adventure*[9] in which two well-educated English women decided to visit Fontainebleau. While walking in the grounds of the summer palace of Louis XVI and Marie-Antoinette in 1901, they saw the grounds near the Trianon as they had been in a much earlier time as determined after extensive archival research conducted over several years. They put the year corresponding to their vision as 1789. The time-shift scales are quite different, but with other examples,[10] it is clear that certain people can experience in waking state the past and the future at a specific geographic site where the present state is suppressed for a short duration. Strikingly, these experiences, like the corresponding dreams, seem to all occur willy-nilly! For example, Eleanor Jourdain, one of the women, spent considerable time in Fontainebleau researching the history of the locale and visited the grounds on her own around the Trianon on several occasions but was never able to re-experience her shared vision of 1901.

The second account[11] involving Goddard occurred in January 1946. He had just arrived at a servicemen's party in Shanghai. Moving through the crowd enjoying drinks, he overheard an officer talking about his dream in which Air Marshal Goddard was killed in a plane crash. Goddard immediately introduced himself and asked the officer for details. Apparently, the crash had been caused by icing on the plane's wings, and there were three passengers—two men and a woman. The crash site was seen to be on a pebbly beach near mountains.

Goddard was scheduled to fly a military version of the Douglas Dakota to Tokyo, and he did not plan on having passengers, so he was somewhat relieved. But by the end of the evening, he had been persuaded to take two men and a woman as passengers. The subsequent flight ended in a forced landing on a pebbly beach on Sado Island, Japan, with a backdrop of mountains. The cause of the downing was icing on the plane's wings. All occupants of the aircraft survived.

Some comments follow.

1. In Goddard's airfield story involving a glimpse of the future (a case of prevision) and the two English women's experience at Fontainebleau from the past (retro-vision), real-life images belonging to another era suddenly appeared over a localized area. These events were exactly registered with the terrain where they had occurred or would later occur. There is apparently no evidence in these two cases that the observers interacted with people they saw moving around.

    At the Trianon in Fontainebleau, the women reported

hearing footsteps and voices. On the face of it, that seems to fit the occurrence of transient information transfer (in holograph form) between universes. It certainly constitutes data to be considered as supporting a time-zoned multiverse. Is that the purpose of these occurrences? Or are they just flaws in the multiverse operation waiting to be exploited as I am doing?

2. For the precognitive dream story involving Goddard, it is a not uncommon situation in which a dreamer sees what appears to be a serious plane crash, in this instance, and immediately assumes that people were killed though such details are not in the dream. In this case, there were no casualties. Notably, four physical details in the dream were correct! This story obviously could have had a *Titanic* effect twist to it, which again increases my opinion that the effect (advocated by some people) is quite unreliable.

Notes

1   J. B. Priestley, *Man and Time* (London: Aldus Books, 1964). I quote from the 1989 edition published by Crescent Books.
2   Later Sir Huw Pyrs Wheldon, OBE, Military Cross, who was at that time running his own BBC program 'Monitor', a major program involving culture and the arts.
3   American Association for the Advancement of Science.
4   This term refers to the scientific establishment, which is a complex of fields each with its own standard model. Any attempted threat to the standard model(s) integrity is viewed with open hostility by the old guard. Change often occurs after they die off!
5   J. B. Priestley, *Man and Time* (London: Aldus Books, 1964). I quote from the 1989 edition published by Crescent Books.
6   G. Ashe, *Encyclopedia of Prophecy* (Santa Barbara, CA: ABC-CLIO, 2001). These statistics show that reliable high performance dreamers of the future (or the past) are a minority amounting to no more than 1 percent of the population.
7   Later Air Marshal Sir Victor Goddard, CBE, KCB.
8   Probably a Hawker Fury.

9   C. A. Moberly and E. Jourdain, *An Adventure* (London: Faber & Faber, 1931). First edition published anonymously in 1911. This 4th edition contains an interpretive note by J. W. Dunne, who did not report having had this type of vision himself. He was not able to explain the adventure, as he continued resisting the idea of multiple universes.

10 For example: S. A. Schwartz, *The secret vaults of time* (Charlottesville, VA: Hampton Roads, 2001). The first edition was published by Grosset & Dunlap, 1978.

11 This was made into a movie film: *The Night My Number Came Up* (1955).

APPENDIX 5

Some "Selfie" Quotes from Time and the Multiverse

These are not of the usual humorous/witty genre; they lean rather toward the philosophical side. In fact, this whole book is not meant to be entertaining unless someone in the physics community shouts out from the balcony: "Hey Mr. Holdsworth! You must be joking!"

1. To understand type-2 time or Whitehead's formal time and our regular clock time practice suppressing it in your writings. An example follows.

2. "Events present and events past
    Are both derived from events future.
    If all events are eternally present
    All events are accessible."

    (This is my rendition of T. S. Eliot's first stanza on time in "Burnt Norton" in "The Four Quartets" in which the word *time* has been eliminated. In the process, it removes one line and corrects an error—according to the interpretation of the STZM model.)

3. Precognition is not as has been claimed an impossibility—but is the result of a naturally occurring window through which we access other universes.

4. In all my dealings with building diagrams the Cosmic Blueprint is unique: it is the most tightly packed with processes—all happening simultaneously—and as well, it has two types of time and two sets of orthogonal axes. That is the next best thing to being impossible!

5. Could it be that we are deliberately positioned nearly midway between the infinitesimally small and the unimaginably large scales of the universe in order that our hyperactive minds and ingeniously tinkering hands are minimized from doing major damage to our planet or even beyond it?

6. The looped nature of this particular multiverse allows for efficient recycling of material after it has been spent at the end of a cycle. It thus represents the largest recycling scheme known to man—and *it reflects intelligence.*

REFERENCES

Barbour, J. *The end of time*. Oxford: Oxford University Press, 2000.

Barrow, J. D. *Living in a Simulated Universe: Universe or Multiverse?* B. Carr, ed. Cambridge: Cambridge University Press, 2009, 481–86.

Bohm, D. *Wholeness and the implicate order*. London: Ark Paperbacks, 1983; first published by Routledge and Kegan Paul, 1980.

—————. *The undivided universe: An ontological interpretation of quantum theory*. London: Routledge, B. Hiley, translated 1993.

Brockman, J. *What we believe but cannot prove*. New York: Harper Perennial, 2006.

Brown, J. R. *Quest for the quantum computer*. New York: Simon & Schuster Touchstone Books, 2001.

Chown, M. *The universe next door: The making of tomorrow's science*. Oxford University Press, 2002.

Davies, P. *About time*. New York: Simon & Schuster, 1995.

Deutsch, D. *The fabric of reality: The science of parallel universes and its implications*. New York: Penguin Books, 1997.

Dossey, L. *The power of premonitions*. New York: Dutton, 2009.

Dunne, J. W. *An Experiment with Time*. London: Faber & Faber, 1927; 3rd ed., 1973.

Flammarion, N. C. *L'inconnu (The Unknown)*. London: Fisher & Unwin, 1900 (first English edition).

Gribbin, J. *In search of the multiverse*. London: Penguin Books, 2010.

Gunn, J. A. *The problem of time*. London: George Allen & Unwin, 1929.

Haisch, B. *The purpose-guided universe: Believing in Einstein, Darwin, and God*. Franklin Lakes, NJ: New Page Books, 2010.

Hammerschmidt, W. W. *Whitehead's philosophy of time*. New York: Russell & Russell, 1947.

Heisenberg, W. *Physics &Philosophy*. New York: Harper-Collins, 1958.

Hunt, V. *Infinite mind: Science of the human vibrations of consciousness*. California: Malibu Publishing, 2nd ed., 1996.

Kafatos, M. and R. Nadeau. *The conscious universe*. New York: Springer-Verlag, 1990.

Krauss, L. M. *I believe that our universe is not unique*, in J. Brockman, ed., *What we believe but cannot prove*. New York: Harper Perennial, 2006, 214–15.

Lloyd, S. *Programming the universe*: *A quantum computer scientist takes on the cosmos*. New York: Knopf, 2006.

Mitchell, E. D. *Psychic exploration: A challenge for science*. J. White, ed. Florida: Capricorn Books, 1976.

Moberly, C. A. and E. Jourdain. *An Adventure*. London: Faber & Faber, 1931.

Myers, F.W.H. *Human personality and its survival of bodily death*. R. Targ, ed. Charlottesville, VA: Hampton Roads, 2001.

Pagels, H. *The Cosmic Code: Quantum Physics as the Language of Nature*. New York: Simon & Schuster, 1982.

Polkinghorne, J. *Beyond Science: the wider human context*. Cambridge: Cambridge University Press 1996.

Powell, D. H. *The ESP Enigma: The scientific case for psychic phenomena*. New York: Walker Books, 2009.

Priestley, J. B. *Man and Time*. London: Aldus Books, 1964.

Radin, D. *The conscious universe*. San Francisco: Harper Edge, 1997.

—————. *Entangled Minds*: *Extrasensory Experiences in a Quantum Reality*. New York: Paraview Pocket Books, 2006.

Richet, C. *Thirty Years of Psychical Research*. New York: Macmillan, 1923.

Schwartz, S. A. *The secret vaults of time.* Charlottesville, VA: Hampton Roads, 2001.

Shermer, M. *In Darwin's shadow: The life and science of Alfred Russel Wallace.* New York: Oxford University Press, 2002.

Smolin, L. *Time reborn.* New York: Houghton Mifflin Harcourt, 2013.

Weiss, B. *Same soul, many bodies.* New York: Free Press, 2004.

Whitehead, A. N. *Adventures of ideas.* Cambridge: Cambridge University Press; New York: Macmillan, 1933; reprinted 1962.

Wolf, F. A. *The spiritual universe.* Needham, MA: Moment Point Press, 1999.

INDEX

16, 17, 20, 21, 22, 23, 24, 26, 29,
30, 31, 32, 34, 35, 36, 37, 38, 39,
40, 43, 44, 46, 52, 56, 67, 68,
69, 76, 77, 78, 80, 83, 87, 91, 93,
94, 96, 97, 98, 99, 101, 102, 103,
104, 105, 106, 107, 108, 120,
123, 124, 125, 126, 127, 128,
132, 133, 134, 135, 137, 138,
139, 140, 145, 149, 150, 151,
152, 153, 156, 157, 160, 162,
163, 164, 165, 166, 167, 168,
169, 170, 172, 173, 175, 177,
178, 180, 181, 183, 184, 185,
186, 187, 188, 189, 190, 192,
194, 195, 196, 198, 200, 201,
202, 205, 208, 216, 232, 235,
236, 237

multiverse computer system 126
Myers, Frederick 140

N

neo-Newtonian 21, 124, 194,
  196, 197
Newton, Isaac 1, 97, 195
Nine-eleven (9/11) 112, 119, 179
non-locality 99, 186
Now (also Now-moment) viii, ix, xiv,
  2, 3, 4, 5, 6, 7, 9, 10, 11, 12, 13,
  14, 17, 19, 20, 22, 23, 29, 32, 33,
  35, 36, 38, 39, 40, 41, 42, 43,
  45, 46, 47, 52, 57, 58, 77, 79, 81,
  82, 88, 94, 95, 98, 99, 101, 102,
  103, 104, 105, 106, 109, 110,
  114, 124, 125, 131, 133, 135,
  136, 137, 138, 140, 143, 147,
  150, 152, 153, 161, 162, 165,
  166, 167, 168, 171, 173, 180,
  186, 188, 190, 193, 197, 200,
  202, 205, 214, 215, 217, 224

O

Osty, Dr Eugene 207, 208, 209

P

Pagels, Dr Heinz 121, 122, 126, 154,
  156, 157, 238
paranormal 7, 30, 31, 48, 61, 72, 87,
  89, 92, 116, 142, 148, 149, 155,
  198, 204, 207, 208, 209, 225
past, present and future 22, 45, 132
Patton, George 34
Pauli, Wolfgang E. xiii, 66, 71, 72,
  73, 74, 75, 76, 77, 79, 80, 81, 82,
  89, 92, 94, 100, 128, 142, 149,
  157, 181, 183, 184, 191, 194, 195
Penrose, Sir Roger 149, 169, 195, 203
Petrie, Flinders 160, 193
Planck, Max 2
Plato 130, 153, 190
Poincaré, Henri 59
Polkinghorne, John 189
precognition viii, xi, xii, xv, 7, 8, 9,
  29, 30, 31, 32, 33, 34, 36, 48, 53,
  54, 63, 66, 87, 88, 89, 92, 94,
  95, 100, 107, 111, 113, 119, 120,
  134, 140, 145, 148, 151, 156,
  161, 163, 166, 169, 185, 186,
  188, 191, 192, 194, 198, 199,
  200, 202, 207, 208, 211, 212,
  215, 220, 229, 230, 236
predetermination 33, 112, 180
premonition viii, x, xii, 7, 9, 29, 30,
  32, 34, 48, 53, 54, 94, 95, 113,
  119, 120, 122, 148, 156, 186,
  192, 211, 212, 213, 215, 220,
  221, 226, 229, 237
Priestley, John B. xii, 54, 87, 223
Puthoff, Harold 127

Q

quantum Mechanics xiv, 2, 6, 26, 55, 69, 74, 75, 76, 77, 81, 82, 88, 100, 123, 124, 125, 127, 128, 151, 166, 168, 170, 173, 186, 187, 194, 198

quantum theory 2, 96, 121, 123, 124, 127, 143, 152, 176, 177, 237

R

Radin, Dean 185
reincarnation 34, 35, 49, 92, 202
retro-causality 33, 169, 173, 175
retro-cognition 30, 31, 33, 48, 151
Richet, Dr. Charles, MD 142, 159
Robertson, Morgan 117
Russell, Bertrand 59, 60, 61
Ryan, Paul 122, 156

S

Scott, Capt. Robert Falcon 212
Scott, (Lady) Kathleen 211, 217, 220, 221
serial time xiii, 2, 23, 51, 65, 67, 68, 69, 70, 78, 80, 91, 94, 97, 101, 120, 134, 160, 166, 168, 181, 182, 183, 187, 196, 227
Serial Time Zoned Multiverse (see also STZM) 23
Seward Icefield vii
Shakespeare, William 139, 141, 159, 216
Shimony, Abner 100
Smith, Capt. Edward 116
Smolin, Lee 151, 197
Society for Psychical Research xiii, 54, 59, 66, 88, 209

soft free will 61, 120, 156, 179, 180, 181
soul xv, 34, 35, 36, 47, 49, 68, 82, 93, 94, 114, 139, 140, 141, 144, 150, 156, 158, 159, 178, 179, 184, 185, 186, 189, 190, 196, 199, 201, 203, 205, 239
Stanford Research International 111
Stead, W. T. x, 113, 115, 118, 219
Stevenson, Ian 49, 117, 155
Sullivan, J. W. 89

T

Taft, William Howard 118
Tegmark, Max 169
Teilhard de Chardin, Pierre 36, 159
Template 8, 9, 17, 19, 24, 25, 41, 54, 60, 61, 68, 76, 77, 86, 87, 93, 98, 104, 107, 113, 126, 127, 129, 132, 137, 139, 140, 144, 159, 160, 167, 168, 175, 178, 179, 180, 181, 183, 184, 192, 201, 202, 219
Terzian, Y. 49
Tesla, Nikola 95, 96
Time Court 108, 109
Time Pips 5, 17, 52, 81, 101, 104, 106, 165, 168
Time step 16
Titanic x, xi, 112, 115, 116, 118, 134, 136, 137, 139, 154, 155, 156, 180, 219, 226, 232
Titanic Effect 112, 139, 156, 180, 226, 232
Toffoli, Tommaso 125, 126, 153
torus manifold 19, 105, 132, 134, 165, 172, 186, 229
Twain, Mark (see Samuel Clemens) xiii, 112, 114, 153

Type 1 time 16, 81, 104, 131
Type 2 time 81, 162

U

Unamuno, Miguel de 189, 191, 199

V

Vanchurin, Vitaly 26

W

Wafer, Lionel 146, 161
Wallace, Alfred 142, 159, 189
Washburn, A. Lincoln ix
Washburn, Land ix
Weinberg, Steven 190, 205, 206
Wells, H. G. 47, 56, 64
Wheeler, John A. 197
Whitehead, Alfred N. xii, 15, 42, 51,
    52, 58, 61, 62, 82, 84, 101, 108,
    109, 129, 130, 131, 132, 133,
    135, 137, 147, 149, 153, 157, 162,
    164, 167, 172, 173, 183, 189,
    235, 238, 239
Wilson, Edward O. 190
Wolf, Fred Alan 140, 158, 189,
    191, 239
Wollheim, Richard 89
Wood, Foresta vii, viii, ix, x
Wood, Walter Abbott vii
wormhole aka ERB 177

Y

Yakutat ix, x
Young, Geoffrey W. 217, 218, 221

Z

Zero Point Field 126, 127
Zimmerman, Dean 31, 44

ABOUT THE AUTHOR

Gerald Holdsworth, PhD, was educated in New Zealand and the United States. For four decades, he researched in both polar regions in the field of glaciology. He's published more than 120 articles in journals and popular magazines and has contributed to several scientific books and government reports. Holdsworth was a research scientist with Environment Canada before working at the Arctic Institute of North America, University of Calgary. He lives in British Columbia.